RODALE PLANS

Insulating Window Shade
Reduces Heat Loss
Through Windows by 80%

by
Ray Wolf

Rodale Plans
Rodale Press
33 East Minor Street
Emmaus, PA 18049

Printed in the United States of America on recycled paper containing a high percentage of de-inked fiber.

Library of Congress Cataloging in Publication Data

Wolf, Ray.
  Insulating window shade.
  (Rodale plans)
  1. Window-shades. 2. Insulation (Heat)
I. Title.
TH2276.W78   690'.182   80-14140
ISBN 0-87857-311-9 paperback
2 4 6 8 10 9 7 5 3 1 paperback

Technical Illustrator
Frank Rohrbach

Carpenter
David Sellers

Project Designer
Dennis Kline

Thermal Engineer
Robert Flower

Book Designer
George Retseck

Copy Editor
Felicia D. Knerr

Cover Photograph
T. L. Gettings

Photographer
John Hamel

Table of Contents

Section I
Introduction

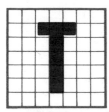his is a book about energy—specifically, how to save it. For many people, windows are the biggest single source of heat loss in the home. If you've already insulated your home, you've cut heat loss through the roof and walls, but what can you do about the windows?

No one wants to board up their windows to reduce heat loss, but how much longer can you go on with heat literally pouring out the windows all night long? The answer is to use insulation that can be put into place when you need it and removed when you want to see out, ventilate, or let the sun shine into the room. That is exactly what Rodale's Insulating Window Shade is. It rolls up out of the way during the day, but pulls down, reducing heat loss, at night.

Although the idea behind the shade is quite simple, the actual design was difficult to finalize. We knew we wanted a shade that would have an R-value of about 5, but it had to be made from common materials, require no special construction techniques, and be attractive. It had to be variable to fit a wide range of windows, and it had to have a cost that was lower than commercially available window insulation systems and had a reasonable pay-back period.

The challenge of developing such a shade fell on Dennis Kline of the Rodale Design Center. Within a few months he came up with the idea of using two rollers and spring-loaded hinges on the side closers. This allowed us to have loose layers of fabric, an unbroken vapor barrier, and a very tight seal on the sides.

We initiated cold-room testing to find the R-value — the resistance to heat loss — of the shade. The shade performed as well as we had calculated it would, and we began the process of putting out a Rodale Plans Book.

This book is different from other types of books. First, it is more than just a book. It is a book, owner's manual, and blueprints, all bound into one. It is designed to be used intimately with the blueprints.

The book is divided into three sections. Section I explains why windows need to be insulated, the different types of heat loss through windows, how much return you can expect from your investment in insulating shades, and the design of the shade. We have put a lot of time and energy into designing the shade, and we are confident that we have developed the best shade possible. Before you make any changes in the design, read why each feature was included in the shade and be sure you understand why it is designed the way it is.

Section II contains detailed instructions on how to measure for your shade, buy and sew the fabric, build the roller box, and install the finished shade.

In many ways, this instructional section is the most unique part of a Rodale Plans Book. It explains each and every step in making a shade. For many, it will be too simplistic; for others, it will mean the difference between success and failure.

Section III comprises the blueprints.

These are not just standard engineer's blueprints; they are specially designed prints to help beginners understand how the shade works. They use simplified drawings and lots of labels and explanations. All steps covered on the blueprints are also explained in section II.

Last, the plans and manuscript were tested. We brought someone not associated with Rodale Press into our shop, gave him a copy of the manuscript, blueprints, and illustrations, and watched him build a shade.

This book is the result of the efforts of many people. Dennis Kline gets credit for the initial shade design. Frank Rohrbach did the illustrations and worked out many of the little bugs of the shade, especially some of the unique retrofit applications. Robert Flower offered constant advice and designed the thermal testing of the shade. Kim Greenawalt and Fran Burghardt worked many hours typing and retyping page after page of manuscript. These people get all the credit, while if any errors have slipped through, they must be considered my fault.

I encourage you to use this book over and over, until every window in your home has an insulating shade. Then get your neighbors to build them. Insulating window shades can be at the heart of an energy conservation revolution in this country, if enough people build them. Let's get the revolution started by building your first insulating window shade.

Ray Wolf

1 WINDOWS AND HOW TO IMPROVE THEM

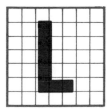et's start right off by pointing out that windows are not the bad guys of the energy picture, as a lot of people would have you believe. They have their drawbacks, but all told, windows are highly desirable. Windows allow us to ventilate our homes, view the outdoors, and even generate heat in the winter through solar energy. The problem is that windows are not always used the way they should be, and they are not very adaptable to changes in climatic conditions.

The ideal window would be insulated with a substance that allowed light to pass through, could be opened, but would give the window the same insulating value as the walls surrounding it. Unfortunately, we know of no such transparent insulation. Thus, windows normally are considered the Achilles' heel of home energy conservation efforts.

For a minute, visualize yourself in a simplified version of your home — four walls, with one window in the middle of each wall. The walls are insulated to an R-value of 19. R-value is a way to gauge a material's resistance to the movement of heat. The higher the R-value, the greater the insulating ability of the material.

OK, here you sit in the middle of your one-room home, with R-19 walls on all four sides of you and an R-30 ceiling over your head. The only problem is that each wall has a big opening in the middle of it, a window, rated at an R-value of 1. You can quickly begin to see where your heat will be going. Heat is somewhat insidious, as it always flows to the coldest spot available.

Windows are nothing more than holes in the thermal envelope the walls of our homes form around us. The goal of a house is to insulate us from outside temperature extremes. This is normally done by installing insulation in the walls and ceilings and caulking all openings to stop air leaks. However, windows have very little insulation value.

If you live in an average home, most likely you are losing from 25 to 35 percent of your heat through the windows, even though in an average home windows account for only 15 percent of the wall surface. If you live in a well-insulated home, with a lot of windows, heat loss through windows may be as high as 50 or even 60 percent of the home's total heat loss.

Look at a south-facing window. For ten hours a day during the dead of winter, that window allows solar energy into your house, warming the house free. East- and west-facing windows do the same, but not to the same degree. North-facing windows allow light but not much solar energy into a home. The ideal situation is for the window to let in the sun during the day, and then at night be covered up and insulated. That requires some form of movable insulation, a relatively new field of energy conservation.

The idea behind movable insulation is that you move it where you want it, when you want it. When you want the sun to shine in, you remove the insulation, but at night when you want to hold the heat in, you put the insulation back into place. There are many types and schemes of movable insulation.

By positioning insulating materials, shades, drapes, shutters, or other devices in front of windows during sunless hours, a great deal of energy and money can be saved. If you live in one of the homes we talked of earlier, where as high as 60 percent of your heat loss occurs through the windows, you can cut your fuel bill by almost half with an effective movable insulation system for your windows. A system like Rodale's Insulating Window Shade has the potential to reduce heat loss through windows by as much as 88 percent, depending on your home.

Movable insulation does not need to be of as high an R-value as the surrounding walls to make a major improvement in the thermal performance of the room. That is because your basic window has a low R-value to start with. An "insulating glass" (double-glazed) window will lose ten times more heat than the average wall and the rate of heat loss for single-glazed windows is double that of the double-glazed window. Just adding a storm window to a single-glazed window can reduce its heat loss by 42 percent. As we'll explain later, adding an insulating shade can yield even bigger savings than storm windows or so-called insulating glass units.

Just adding additional layers of glass is not the answer. Every layer of glass you add to a window cuts down on the amount of solar energy that passes through the window. Adding glass is an expensive and inefficient way to reduce window heat loss. Some type of insulating material that is completely out of the window during the day and completely covers it at night is the answer.

Insulating your ceiling and walls makes your windows even more of an energy weak point, and insulating window shades become an even better energy conservation "buy." The final ranking of where your energy conservation dollars will be most effectively spent will differ according to each home, its construction and location, and your lifestyle. But consider the fact that every day one and one-half times the amount of energy delivered by the Alaskan pipeline is lost through windows in America, and you can see that insulating window shades can be important energy savers.

HOW WINDOWS LOSE HEAT

Let's look at the four ways windows contribute to high fuel bills, then see what can be done to take advantage of a window's good points and improve the weak points.

Windows lose heat in four ways: through infiltration, conduction, convection, and radiation. Infiltration is the actual movement of cold air into your house, while the other three types of heat loss will cause transmission of heat out of your house, through the windows in particular. Remember that heat always travels from places of higher temperature to places of lower temperature and always takes the path of least resistance. All four types of heat loss must be reduced if overall thermal performance of a window is to be improved.

INFILTRATION

Let's look first at the most insidious type of heat loss through windows, infiltration. This is the movement of cold, dry, outside air into the house. This infiltration is accompanied by the loss of warm, moist, inside air to the outside. Illustration 1-1 shows how air will infiltrate a typical window. Infiltration is a thief, because your heating system must supply the energy to heat the incoming cold air, plus the energy to maintain a comfortable humidity. Infiltration from all sources is responsible for one-quarter to one-half of the heat loss of a typical house. Even in warm climates infiltration works against you, because it brings in hot air from the outside

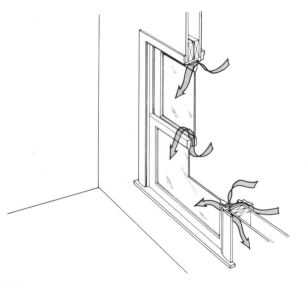

Illustration 1-1 — Air infiltration takes place through any and all cracks around a window.

which displaces cool air inside the house.

Infiltration is a mechanical problem, in that it can be stopped through the use of physical barriers. Caulking and weather stripping will block the movement of air, thus stopping infiltration. Even with the world's best insulating shade installed, caulking and weatherizing your window are the first things you should do.

Once you've solved the physical problem of stopping or reducing infiltration, turn your attention to the three ways heat moves.

CONDUCTION

The only way heat can go through solid objects is by conduction. A small amount of radiant heat will pass out a window, but basically, windows should be considered as solid objects for the sake of discussing heat loss. Since the glass and frames of your windows are all quite solid, any heat leaving your windows must be conducted through these materials at some point or another. Heat reaches the inside glass surfaces in several ways and leaves the outside surfaces in several ways, but it moves through the glass itself in only one way — conduction.

A close-up look at conduction within a solid object would show that the heat moves because the molecules in a hot area move much faster than those in a cold area. When you grab the handle on an iron frying pan and burn your hand, it is because of the conduction of heat from the bottom of the pan on the stove burner to the handle.

Some materials conduct heat much more efficiently than others. For instance, aluminum, copper, and brass are all very efficient at conducting heat, thus we see a lot of aluminum- and copper-bottom cookware. Other materials, called insulators, conduct heat very poorly. Wood is a poor conductor of heat, thus we see a lot of cookware with wooden handles. Glass is a good conductor of heat, as shown by the amount of glass cookware being sold.

Air is a very poor conductor of heat. Most types of insulation are nothing more than pockets of trapped air. Every object has a very thin layer of still air along its surface, serving as an insulator. In a typical window, these invisible air films along the glass surfaces are actually better insulators than the glass itself and are largely responsible for what little R-value the window does have. If these thin surface layers are disturbed, the rate of heat loss through the window will be increased considerably. This often happens on windy days.

To reduce conduction of heat through a window, you have to cover the glass and frame with a good insulating material, to stop the heat from actually reaching the window. Thus, the insulating value of the fabric you use in the shade will help to prevent heat from being conducted as rapidly to the surfaces of the window and on to the outside.

Insulating windows are nothing but two layers of glass with a very small air gap between them. An interesting point concerning these windows is that most of the insulating value comes from the added air films on the extra piece of glass. The actual gap between the window panes does not offer much protection, and the frames and sash are no better insulated than ordinary windows. Considering the high cost of these windows, you are better off with a lower-cost storm window or a good-fitting insulating shade. The best feature of these windows is their new construction which generally reduces air infiltration considerably. For existing homes, the best approach is to caulk the window well and add an insulating shade, rather than to replace the windows or add a storm window.

CONVECTION

The second way heat travels is known as convection. This is actually the motion of the air carrying the heat. Warm molecules move faster than cool molecules, and expand, become lighter, and rise, while the slower, cooler molecules tend to fall. Thus, warm air rises and is normally trapped against your ceiling, while cool air settles on the floor. However, when the warm air eventually cools, it becomes heavier and falls. This movement sets up air flows, known as convection currents, in a room, as shown in illustration 1-2.

Normally, warm air gathers along the ceiling and, as it cools, falls along the face of the window. There it does two things. First it gives off some of its heat to the window through conduction, but, more important, the movement of the air disturbs the air film on the surface of the glass, increasing the conduction of heat through the window.

Convection currents will normally form behind standard drapes. Many of the "insulating" drapes sold today are almost worthless, as convection currents travel between the drape and the window, pulling warm room air into the space at the ceiling, and eventually sending the heat out the window and returning chilled air at the floor.

To prevent convection currents from forming, you need insulation that is sealed on all edges of the window. The insulation will stop the movement of heat from room air to the glass through conduction, and a tight seal will stop the flow of room air to the glass behind the insulation.

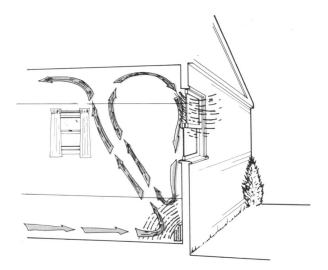

Illustration 1-2 — Convective heat flow is the air pattern created by heat and the way it moves. Warmer air rises and settles as it cools. Falling against a cold window rapidly cools warm air.

An added bonus of stopping convective air currents from forming along the windows is that the room will be much more comfortable to sit in. Convective currents are literally thermal drafts, moving cold air across your body and wicking away your body's heat, eventually carrying it to the colder surface of the window where the heat is lost to the outside through conduction. Stopping convective currents reduces drafts.

RADIATION

The last type of heat movement is thermal radiation, as shown in illustration 1-3. This is a very important type of heat loss, especially for those who heat with radiant wood, coal, or electric heaters. All objects emit radiant energy which travels, like light, through the air until it strikes a solid object. Warmer bodies emit more radiant heat than cooler bodies. A stove warms you because it is warmer than you, and thus emits radiant heat to you, warming your body.

Radiant heat loss is not very hard to understand. Perhaps the best way to understand what is happening is to hold your hand in front of a heat source, preferably a wood stove. You'll feel heat on the side of your hand facing the stove. That is because the radiant energy coming from the stove is being absorbed by your hand. But you don't feel any heat on the back of your hand, because radiant heat, like light waves, travels in a straight line. Now, take a piece of aluminum foil and hold it in front of the stove. Put your hand between the stove and the foil, and

you'll soon feel heat on both sides of your hand. The shiny foil does not absorb radiant heat as your hand does; it reflects it or bounces it back towards the stove.

Radiant heat loss can account for up to 65 percent of the heat lost through a window. That means that the quickest way to improve a window's thermal performance is to stop radiant heat loss. Luckily, radiant heat loss is fairly easy to stop. All you have to do is hang a highly reflective surface in front of the window, and you've cut your heat loss through the window by about one-half.

In fact, a single layer of reflective material is the cheapest type of window insulation you can have. A single reflective layer would not do anything about conductive heat loss, but it would almost totally stop radiant heat loss and would also stop convective heat loss if the edges were sealed. Without a reflective barrier, heat radiates to the glass and is

Illustration 1-3 — Radiant heat loss is the direct traveling of heat to a cooler window. Some radiant heat passes directly out the window, while most is absorbed by the window and lost through conduction to the outside.

absorbed, warming the glass. This heat is then conducted from the room side of the glass to the cold side of the glass, and on to the outside environment.

If you want to see how much radiant heat loss your home suffers, make this simple test. After the sun sets, turn on one light in every room. Walk outside and see how much light is escaping. Remember, radiant heat travels the same as light; if you see light streaming out a window, you are also losing radiant heat. Put a reflective barrier over a window and return outside. How much light do you see now? That is the way to stop radiant heat loss.

In our thermal testing of the shade, we tested a single piece of Mylar and found it to have an R-value of 3.2, more than half the total R-value of the shade. This is based on the amount of radiation and convective heat it prevents from leaving the room. Any type of movable insulation should have a piece of reflective material properly positioned to be really effective. As you'll see in chapter 2, "Design of Rodale's Insulating Window Shade," positioning the Mylar so that other solid materials are not in contact with the reflective surface is necessary to maintain maximum efficiency.

HOW TO IMPROVE WINDOWS

Now that you know all the evil things a window can do to your energy use, let's look at the more positive side of things. Windows can indeed make you money, if they are situated and you use them properly.

During winter months when the sun shines through a window, that is free energy you are collecting. The sunshine will heat at least part of your room free. Some of that heat will be stored by items in the room and will warm you later in the night. A passive solar home works completely on that principle; let enough sun in during the day, and find a way to store it, and it will keep you warm all night — if you can find a way to prevent it from being lost.

For existing homes, the question becomes how efficient can you make your windows? Chart 1-1 shows the calculated yearly savings from a Rodale Insulating Window Shade installed on a window in New York City. The chart is based on a window with a surface area of 10 square feet. A plus value on the chart means that the solar heat gain during the day more than makes up for nighttime heat loss from the window.

You can see that just the plain window will give a small amount of free energy when facing directly south. The east and west windows each show a net loss of slightly more than $3, while the north runs a loss of almost $9 per season.

However, add an insulating window shade to the same window, and look at the financial turn-around. The south-facing window now produces more than $12 per year in energy. That is more than one dollar for every square foot of window. The east and west windows generate almost $5 per window, and

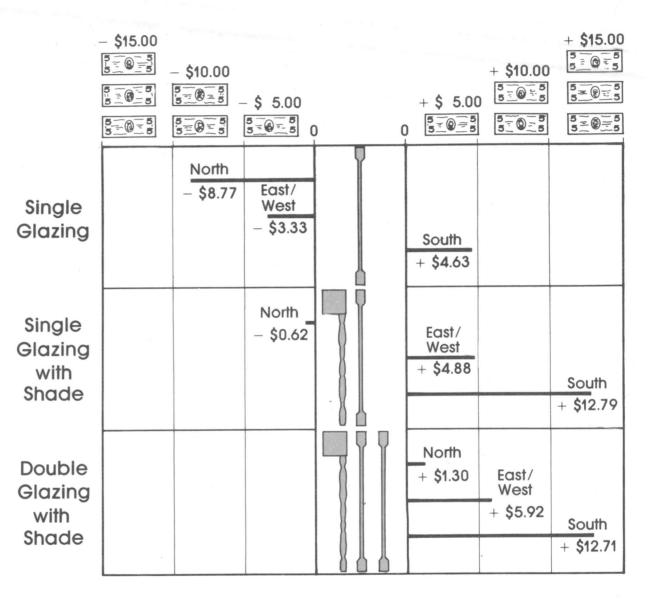

Chart 1-1 — Economics of Installing a Shade on Different-Facing Windows Based on a 10-square-foot window in New York City, with a fuel oil cost of $1.00 per gallon and a furnace efficiency of 60 percent.

the north window loss is cut almost to a break-even point.

The savings when the shade is put on a double-glazed window are even slightly better. An interesting point to note is that the savings go down slightly for a double-glazed, south-facing window, but up considerably for a north-facing window. That is because the second layer of glazing reduces the amount of sun that passes through the window, in addition to reducing the heat loss at night. For the south-facing window, you don't need the extra layer of glazing, and for the east- and west-facing windows, it is questionable whether the extra cost of the second glazing layer is worth the small amount of savings generated. However, for the north window, the extra glazing does not cost you anything in energy gain, yet helps reduce energy loss, so it is a good investment.

Chart 1-1 is based on an energy cost of $1 per gallon for # 2 fuel oil, and a furnace efficiency of 60 percent. Obviously, the higher your energy cost, or the more severe your heating season, the more your savings would be.

You can use chart 1-1 to figure the savings you would get by adding insulating shades to your windows. For instance, the difference made by adding a shade to a single-glazed, south-facing window comes out to a total improvement of $8.16. An east- or west-facing window has a total improvement of $8.21, and a north-facing window, a savings of $8.14. You can see that all three figures are very similar. That shows that the shade will reduce heat loss at the same rate,

regardless of which direction of the compass the window faces.

Chart 1-2 shows the heat savings of several window treatments calculated from our thermal test results. The basic, single-glazed window is considered the baseline, with no heat savings. If you add a storm window to the outside, you reduce heat loss by 42 percent.

If you take off the storm window and add a single layer of insulated fabric, sealed on all sides, to the inside of the single-glazed window, the heat loss is cut by 67 percent. However, a single layer of Mylar reflective film reduces heat loss by 71 percent. But the total insulating shade, with an R-value of 5, reduces heat loss by 81 percent.

How much you will save by installing insulating shades in your house is hard to calculate exactly, but you can do a rough estimate as follows. Our calculations showed that the shade will save about $1 per year in fuel oil for every square foot of window surface covered by a shade, based on an average insulated home and an oil price of $1 per gallon. Just multiply the number of square feet of windows you have by $1 to find the rough savings you can expect. The figure you get will tell you how much you can expect to save in one year.

If you heat with electricity, each square foot of shade will save about $1.25 per year, while if you heat with wood, coal, or gas, each square foot of shade will save you about 65 cents a year. But, regardless of the fuel, your savings will go up automatically as energy prices increase.

If you live in areas that require air conditioning, you will also have an additional savings from the amount of heat gain the shade prevents. This is a lot harder to estimate. Depending on your house, the window location, and other variables, the dollar savings from reduced air conditioner use will vary greatly.

Depending on the type of materials you buy and the size of your window, you should figure that Rodale's Insulating Window Shade will cost about $2.25 per square foot to build. So in an oil-heated home and at a stable energy cost, each shade will pay for itself in just about two years. A shade should last from seven to ten years before you have to replace the fabric, so you can count on a savings for at least five years before the shade needs any attention.

The obvious question is what window will save you the most money by having a shade on it. There is a simple answer. If all your windows are in similar condition, then start putting shades on the largest areas of glass you have. Shades for larger windows have a lower cost per square foot than smaller shades, because the hardware is averaged over a larger area. Plus large expanses of glass are proportionately bigger losers of energy than small windows. Small windows have a higher percentage of trim and molding, which does not lose heat as fast as solid glass.

We have put shades on windows as wide as 7 feet, even on sliding glass doors. The only limiting factor is the length of wood available to make the rollers. The mechanics of the shade will work on large or small windows. If all your windows are in equally good condition, we recommend you install the shade on the largest windows first and work your way down to the smaller windows. As you saw in chart 1-2, the energy savings are the same, no matter which direction the window faces.

However, if you have an especially bad window, one that is very close to your stove or thermostat or has a great deal of infiltration, by all means put the shade on that window first. It does not matter which direction the window faces — if it is in bad shape, put the shade on it first.

One other possibility you may want to consider is to put the shade on those windows you use the most. The shade is a very convenient device to open and close, so put it where you will use it a lot. You may find it better to cover out-of-the-way, north-facing windows with a less expensive, less convenient form of window insulation.

The idea is carefully to pick and choose an insulation idea that is best for your circumstances. We feel that Rodale's Insulating Window Shade is the right choice for about 80 percent of the windows in America. The other windows should have some other type of insulation or none at all.

To better understand how the shade works, go to the next chapter and see how it was designed and how it can greatly improve the thermal qualities of your windows.

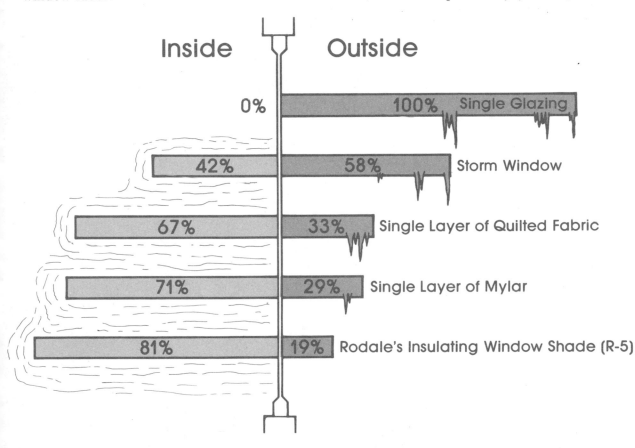

Inside	Outside	
0%	100%	Single Glazing
42%	58%	Storm Window
67%	33%	Single Layer of Quilted Fabric
71%	29%	Single Layer of Mylar
81%	19%	Rodale's Insulating Window Shade (R-5)

Chart 1-2 — Percentage of Heat Saved by Different Window Treatments
Actual test results of different window treatments, based on a single-glazed window in an insulated home.

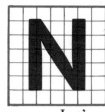ow that you know how and why windows cost you money by wasting energy, it's time to see how Rodale's Insulating Window Shade saves you money by saving energy. Let's go over each design feature of the shade to get a complete understanding of why it is designed the way it is.

INSULATION VALUE

When we set out to design the shade, the main thing we wanted to do was insulate the window. For this, we knew we would need a material with a high R-value. R-value is an indicator of a material's resistance to heat loss. The higher the R-value, the more a material will resist the movement of heat through it. However, we also knew we wanted to be able to roll the shade out of the way when it's not needed. To do this, we obviously needed a flexible material.

We began to look at fabrics that are commonly available, and at their potential resistance to heat loss. We quickly narrowed our search down to a variety of quilted fabrics. These fabrics are made with multiple layers of fabric with an inner lining material. Trapped air is a very good insulator, and in a layered material, such as the quilted fabrics, a layer of air is trapped on every surface. Quilted material has a number of surfaces, thus a number of layers of trapped air and a higher R-value than a single layer of thick fabric.

Even though your walls and ceiling may be well insulated, we found that the level of insulation in a window shade need not be nearly as high as that of a wall. Of course, the higher the R-value of a shade, the more heat it will stop. However, there is a point at which the cost of providing a higher R-value in a movable shade just can't be recovered by the small amount of additional energy savings

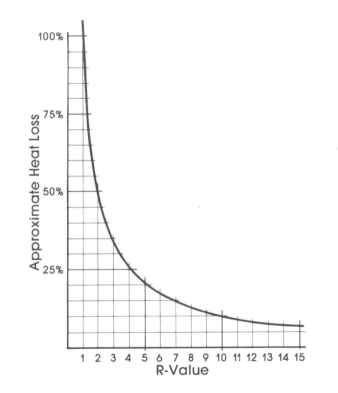

Chart 2-1 – Diminishing Return of Higher R-Value. Initial efforts provide the largest savings when adding R-value to a window.

it provides. Chart 2-1 shows that an R-value of 5 is that point. You can see that the additional percentage of heat saved by a shade drops drastically as the shade's R-value increases beyond that point. This gives you substantial heat savings, yet can be reached without undue cost. We have made some shades using a very thick fabric, giving an R-value as high as 8, and it reduced heat loss through a single-glazed window by 88 percent. However, an R-5 shade reduces heat loss by 80 percent.

The mechanics of building a shade with an excessive R-value are both tricky and expensive. We feel that R-5 is the ideal amount of insulation for a shade. To get a higher R-value would require additional materials and add greatly to the cost. The energy savings would not justify this additional expense.

INFILTRATION BARRIER

The only problem we knew of with using quilted fabric was that the shade would not be a very good infiltration barrier. Air easily moves through the many stitching holes in quilted material, making it almost, but not quite, as porous to air and heat movement as window screens.

Not only does air movement through a shade reduce the thermal effectiveness of the shade, but as warm room air moves through the shade and gathers next to the window, moisture condenses on the cold windowpane. To prevent condensation, we knew we would have to come up with a barrier that not only stopped air infiltration, but served as a vapor barrier as well. We knew that one very good, commercially available insulating shade used an ultrasonic process to bond a fabric layer to a vapor barrier without breaking the surface of the vapor barrier, but we could not expect homeowners to have access to such a process.

Finally we hit on the idea of having two separate layers of quilted material with a third, unbroken, middle layer to serve as an infiltration and vapor barrier. This would give us the best of all worlds — inexpensive and readily available fabric choices and a very good infiltration barrier — all at a low cost.

We then realized that, if this unbroken barrier were highly reflective, it would also serve as a barrier to radiant heat loss. As you saw in chapter 1, "Windows and How to Improve Them," radiant heat loss normally accounts for the majority of the total heat loss through a window. We decided to use a material known as aluminized Mylar, a product of the space race in the 1960s. Mylar is a tough plastic film that is somewhat flexible, yet is totally impervious to water and air. The aluminized coating gives it a very high reflectivity rating. Such a material would serve all three duties we were asking of it. It would stop air infiltration, serve as a vapor barrier, and block radiant heat loss.

Thus we ended up with a shade made of two layers of quilted material with a Mylar vapor barrier in the middle. This gave us a finished product that effectively blocks all types of heat loss through windows. We had the insulating value we wanted, and the

shade stopped radiant heat loss and infiltration and greatly reduced moisture condensation problems on the window, avoiding any possible damage to the windowsill over time. The fact that all the materials were commercially available at reasonable cost made it look like a sure winner.

The only problem was that we had to put all the layers together without reducing the thermal properties of our sandwich construction. If they were simply sewn together, we would not only lose the vapor and infiltration barrier, but the shade would not be as effective at stopping radiant heat loss.

During one series of R-value testing we found that, if the Mylar were in contact with the fabric, the shade did not perform as well. We researched and found that, to gain the maximum reflective value from the material, it was best not to have the Mylar touching the fabric at all. If the Mylar and the fabric touch, heat travels through the Mylar by conduction, and the reflective surface never has a chance to bounce radiant heat back into the room, as shown in illustration 2-1.

In our test we found that the shade had an R-value of less than 5 with the Mylar in direct contact with the fabric. But if we separated the fabric and the Mylar by a ½-inch air gap, the R-value went up to almost 7. Thus a new design challenge was added; finding a way to put three layers together without the Mylar being in contact with the fabric. With the two separated, heat would be radiated to the face of the fabric, travel through the fabric by conduction, and radiate off the other side of the fabric to the Mylar.

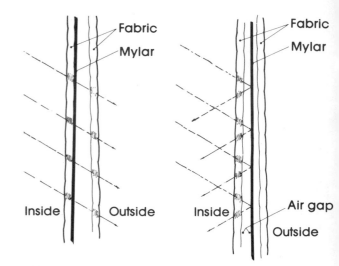

Illustration 2-1 — With no air gap between the Mylar and the fabric, heat is lost by conduction through the Mylar as shown at left. But with an air gap, as shown at right, the reflective properties of the Mylar bounce heat back into the room.

The Mylar would then bounce the heat back to the fabric, which would in turn radiate the heat to the room, as shown in illustration 2-2. If the fabric and Mylar touched, much of this advantage would be lost.

The solution to the problem was to use two rollers, or dowel rods. That way, each layer of fabric would have its own roller, and the sandwich of materials would not need to be sewn together. The Mylar could then be positioned between the two layers.

We ended up with a box, containing two dowel rods, above the window. A piece of fabric would be attached to each dowel rod with standard fabric snaps, and the piece of

Mylar would be attached to one of the pieces of fabric the same way.

Initially we were sewing a loop on one end of the fabric, slipping the dowel rod through the loop, and then stapling or nailing the fabric to the dowel rod. This worked fine, but the sewing was somewhat exacting, and the shade was hard to take off the dowel rod to wash. We found something known as a snap fastener stud. This is nothing more than half a snap with a screw attached. The screw fastens into the dowel, and the snap is attached. These are used mostly for marine applications, but hardware stores carry them.

For the shade they work perfectly. All it takes is four or five snaps, and the fabric is firmly fastened to the dowel. However, it can be easily unsnapped for washing with no problem. An added bonus of using the snaps is that this makes the sewing of the shade extremely simple. All you have to do is cut a rectangle of fabric and edge it.

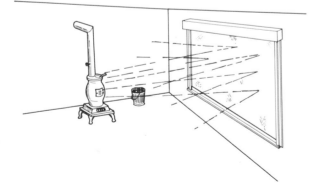

Illustration 2-2 — With a shade in place, radiant heat is bounced back into the room, not lost out the window.

At first we attached the Mylar to the piece of fabric closer to the outside, so it would not be in direct contact with the front piece of fabric. However, we later changed that design so the shade could be reversible.

SUMMER USE

Now the shade has a winter and summer mode. See blueprint sheet 6 for details on how the two modes function. The difference is that in winter the Mylar is between the two layers of fabric, and reflects radiant heat back into the room while serving as a vapor/infiltration barrier. However, in the summer, the Mylar is moved to the outside of the shade where it serves as a barrier to radiant heat from the sun. In this mode it prevents heat buildup in your house.

In fact, if you have double-hung windows, you can open the top section several inches, pull the shade down to within several inches of the sill, and the shade will actually cool your house, as shown in illustration 2-3. As sun comes through the window, it hits the Mylar and is reflected back out the window. This creates a very minor heat buildup between the shade and the window. As the slightly warmer air rises, it goes out the opening at the top of the window. This creates an air pull at the opening at the bottom of the shade, and room air is pulled into the cavity between the shade and the window. As air is heated, it rises and exits out the top. Thus, the window/shade combination is constantly pulling room air in and exhausting it out the top of the window. If you

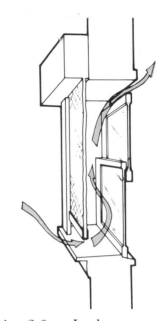

Illustration 2-3 — In the summer, room air will be pulled in at the bottom of the shade, heated, and exited out the top of the window, helping to cool the house.

do this on all your south- and west-facing windows and open your north- and east-facing windows, cool air will be pulled into and through your house, cooling it without using any energy other than the sun.

But, by making the shade reversible, we were faced with the problem of having too many snaps on the outside layer of fabric for the shade to roll evenly. Finally we decided to compromise and put the Mylar on the inside of the room-side layer of fabric in the winter and the outside of the window-side layer in the summer. This balanced the placement of the snaps, and the shade rolled evenly.

BOTTOM MOLDING

However, to keep the Mylar from coming in contact with the inside piece of fabric, we developed a unique bottom molding arrangement. We used a very thin piece of molding on the outside of each piece of fabric, with a thicker piece of molding between the two pieces of fabric. This puts air space between the fabric layers, allowing the Mylar to hang somewhat freely between the layers during the winter months. Some parts of the Mylar are in touch with the fabric, but not all of it. The shade is not as absolutely effective as it could be, but it is better than having the Mylar in contact with the fabric. We felt it was a good compromise. To design it otherwise would have involved more complicated construction methods, and we felt it important to keep the shade as simple to build as possible.

While working with the bottom molding, we looked for ways to seal the bottom against the windowsill. We tried using standard weather stripping but decided against it in favor of letting the fabric of the shade become the seal. By making a 1-inch double hem in the bottom of the shade and leaving at least ¼ inch of this hem hang below the bottom molding for each piece of shade, we created an effective double bottom seal. When you pull the shade down, push the bottom firmly against the windowsill before closing the side closers, to seal the bottom.

With the shade itself finally designed, we needed to find a way to roll it up and down. Our method uses a plywood pulley on each dowel rod, with one pull cord. The pulleys are made of three separate pieces of plywood glued together.

Each finished pulley has a couple of wide rubber bands placed over the center of the pulley to give added traction to the pull cord. We used standard drapery pull cord, and it works just fine. Some of the newer polyester cord products are more slippery than the old cotton rope cords and will not work as well.

The pull cord must be formed into an endless loop. To do that, we tried many devices and finally settled on a standard electrical butt splice. These are designed for butting two wires together. An uninsulated electrical butt splice for # 10 – 12 wire fits drapery cord perfectly. All you need to do is cut the cord to length and insert each end of the cord into an end of the splice, crimp the splice tight over the cord, and you've formed an endless loop of cord.

Blueprint sheet 6 shows how the cord must be wound around the two pulleys for the shade to work. It is mandatory that the cord be wound correctly in order to drive the shade's dowel rods in the right direction to roll the shade up properly.

The last thing we needed was a way to keep the shade from rolling down whenever the side closers were opened. For this, we found that keeping constant tension on the pull cord worked well. To do that, we use a simple drapery cord tensioner. This is nothing more than a spring-loaded pulley that mounts on the wall below the windowsill. The pull cord passes through the tensioner, and the spring keeps constant tension on the cord. The cord thus keeps the shade in place as you pull it into position. The only problem with the tensioner is that the butt splice to connect the pull cord will not pass through most tensioners. This will create occasional operating problems for the shade, as explained in chapter 7, "Installing the Shade," and blueprint sheet 6. This minor problem can be overcome with careful operation of the shade or an occasional readjustment of the position of the splice.

MOUNTING

With the mechanics of the shade all worked out, it became a question of how to mount the shade. We first thought of making the shade part of the actual window. However, when we read a study pointing out that 60 percent of air infiltration through a window normally moves around the edges of the window trim, we changed our minds. We started to look at ways to mount the shade flat on the wall, surrounding the existing window, as shown in illustration 2-4.

Finally we decided to use the window trim itself as part of our sealing mechanism. By mounting the roller box above the window and making the fabric the exact width of the window trim, the shade would seal out all air infiltration, and we would not have to get involved in removing window trim or even putting nails or screws into the trim.

We have calculated the shade design and the data sheet so that the box will overhang the window evenly on each side,

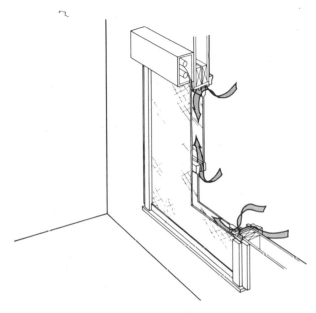

Illustration 2-4 — Air infiltration is prevented from getting into the room from behind the shade.

and the shade will completely cover the window and the side trim, as shown in illustration 2-5. By adding caulking when installing the roller box and side units, you can almost totally seal off the window.

The last question we had to consider was how to seal the sides. We tried a number of metal and plastic channels, but none worked well enough. The problem was always the same; you want a very tight edge seal, but if you get too tight a seal, the fabric will not move up and down smoothly. All commercial shades make a compromise in this area. They have an edge seal that loosely cradles the fabric, yet allows it to move without placing undue wear and tear on the material.

We finally decided on movable side closers, something that could be loosened to position the shade and tightened when the shade was in place. We tried making the sides out of wood, using a cam-type action, but that proved somewhat difficult. The eventual answer proved to be the simplest of all approaches. We used spring-loaded cabinet hinges to seal the sides. These hinges exert pressure when closed, yet can be opened to relieve all pressure against the fabric. Thus, when the hinges are opened, the fabric moves freely with no danger of wear and tear. This will add greatly to the life expectancy of the fabric. These hinges are readily available, easy to install, and have a low cost.

We felt that the shade was then complete. We had an easy-to-build, low-cost unit that performed well. However, when we tested it, several minor problems surfaced. Whenever the top was opened, the dowel rods fell out. To correct this, we changed the shape of the rod holders, so that a dowel rod is now cradled in an opening and will not fall out when the box is opened. To do this requires a somewhat tricky cut, but blueprint sheets 3 and 4 detail exactly how to lay the piece out to avoid any errors.

But the biggest problem we found with the shade was that it only worked really well on one type of window. It was designed for mounting flush on the wall. This is fine for a lot of windows, but for windows with deep windowsills, it created problems. If you have plants in the window and pull the shade over them at night, chances are good they'll freeze

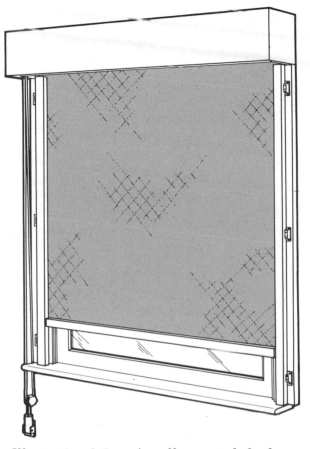

Illustration 2-5 — A wall-mounted shade covers an entire window opening.

on a cold night. With a large air gap between the shade and the glass, the shade loses some of its effectiveness, and you could have problems with condensation.

Thanks to the vapor barrier, no moisture moves into the area between the shade and the glass. However, with deep windowsills, a lot of moist air is trapped between the shade and the glass when the shade is pulled into place. When the volume of air is too great between the shade and window, convective currents will be started, as shown in illustration 2-6, and moisture can create problems by condensing on the window and damaging the woodwork.

The solution we hit on was to mount the shade as close as possible to the glass of the window. To do this, we redesigned the shade for a second mounting option. We call the second design the window-mounted shade, and it is designed to fit into windows where the sides of the roller box will not show, as shown in illustration 2-7. To do this, we had to redesign several pieces of the roller box. The shade works the same as the wall-mounted one but is assembled differently.

In section II you'll see that each type of shade is explained separately. The window-mounted shade uses different types of side closers and is not quite as effective at reducing infiltration as the wall-mounted shade.

Chart 2-2 shows heat loss by conduction and convection. You can see that their combined effect is minimized at an air space of about ¾ inch. As the space between the shade and glass increases, convection currents become established between the shade and the glass, as shown in illustration 2-6. This leads to increased conduction from the shade and increased moisture condensation.

Last, we felt that having the roller box mounted inside a deep window made for a much nicer looking installation. This way the box is almost totally hidden and is only noticed at night when the shade is down and saving energy.

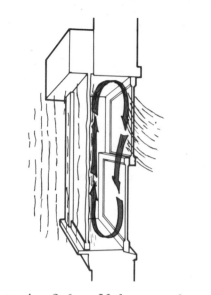

Illustration 2-6 — If the space between the shade and the glass gets too large, convective currents will become established, leading to increased heat loss and added moisture condensation.

We haven't been able to make the shade perfect. The major drawback to the shade is that, because it is such an effective insulation device, it darkens a room when the fabric is pulled down. No light filters through the shade at all. In effect, it is a type of blackout shade, as it is lightproof and sealed on all four sides. The only way around this problem is to create a new, transparent type of insulation that does all the shade does, yet allows light to pass through. It is doubtful that anyone will be able to do this at a reasonable price in the foreseeable future, although lots of companies are trying.

The only other problem the shade has is that the pull cord must be pulled differently, depending on what season the shade is set for. Blueprint sheet 6 explains this fully. The shade's rollers must revolve in certain ways for the fabric to roll up smoothly. Thus you have to pull one side of the cord in winter and the other side of the cord in summer. Occasionally the cord will slip, and the butt splice holding the two ends of the cord together will get caught in the tensioner. The splice will not travel through the tensioner. When this happens you have to remove tension from the cord, open the box, and reposition the splice. This will not happen very often, especially if you are careful not to let the shade fall into

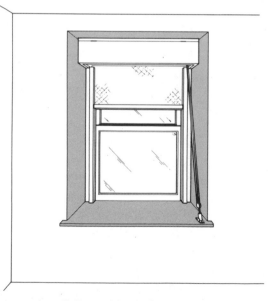

Illustration 2-7 — A window-mounted shade goes in the window opening, and mounts almost directly against the glass.

position. Gently pull it into position, and close the side closers. If you let the shade fall out of the roller box like a guillotine, there will be more cord slippage and the splice will get in the way of the tensioner more often.

Besides those two small problems, we know of nothing that could be done to improve the performance of the shade. It is easy

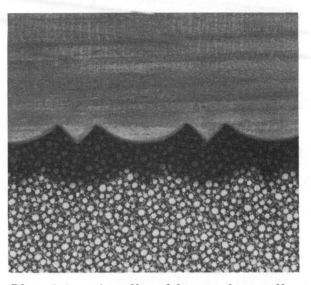

Photo 2-1 — A scalloped front makes a roller box much more attractive. Such a front requires no changes in the size of the pieces, only careful cutting.

to build, inexpensive, and effective.

There are a great many things you can do to make your shade and roller box attractive. Photo 2-1 shows a unique treatment of the face of the roller box to give a somewhat more homey feel to the shade. The shade fabric can be chosen to enhance a number of interior design features. Do whatever you want to make the roller box look the way you want it to. Wallpaper it or cover it with fabric — anything goes.

All that remains between you and your finished shade is section II, the step-by-step instructions for building your first insulating window shade.

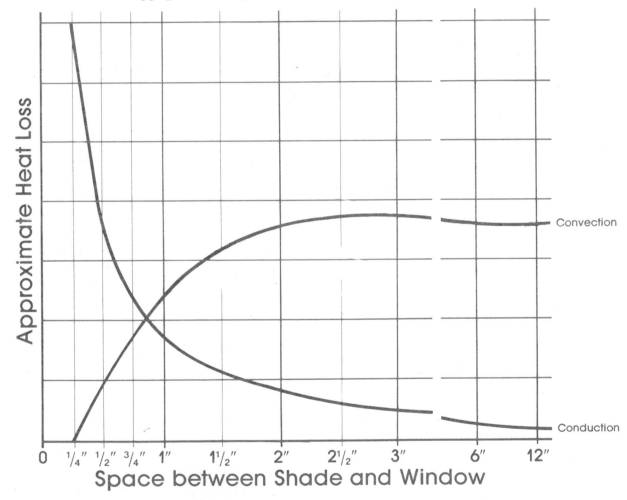

Approximate Heat Loss

Convection

Conduction

| 0 | ¼″ | ½″ | ¾″ | 1″ | 1½″ | 2″ | 2½″ | 3″ | 6″ | 12″ |

Space between Shade and Window

Chart 2-2 — Comparison of Conductive and Convective Heat Loss
As the gap between shade and the window increases, convective heat loss increases. At close spacing, conductive is the major form of heat loss; with a larger gap, convective air currents become established.

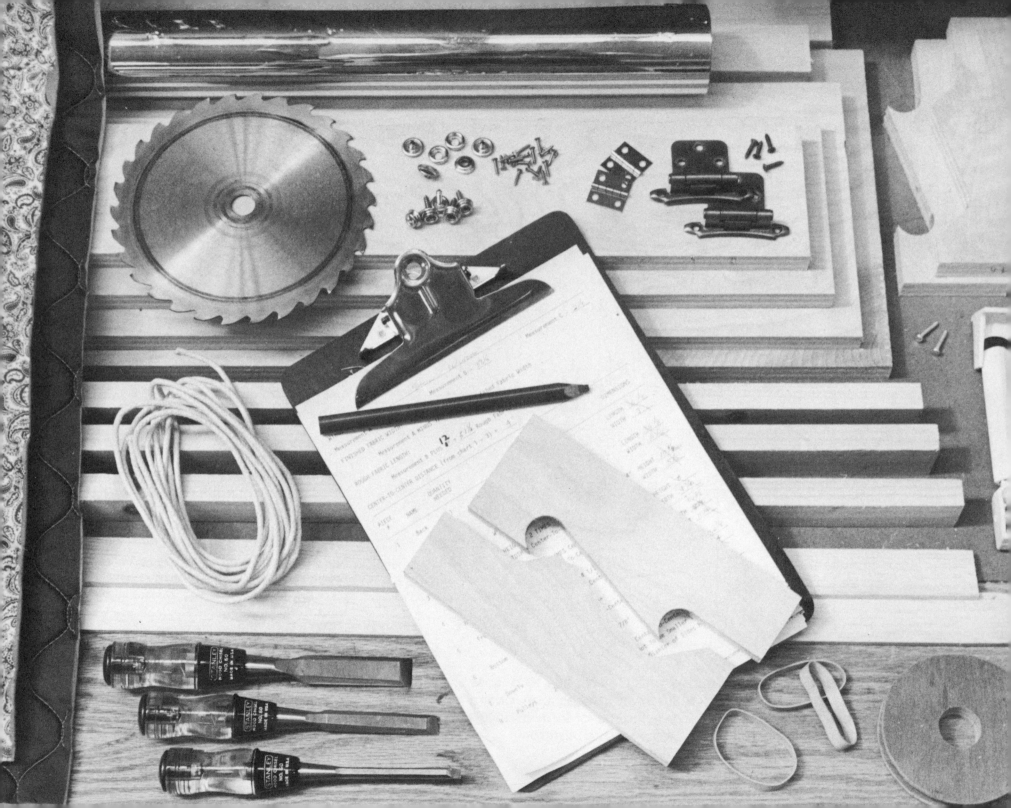

Section II Introduction

This is the nuts-and-bolts part of the book. It's time to get out your tape measure and tools, buy some lumber and fabric, and start making your insulating shades.

This part of the book is designed to enable even rank beginners to build an insulating window shade successfully. For some, the instructions will be too simple; others will be taxed to their limit. Use the instructions and illustrations in conjunction with the blueprints to get over any tough spots. The blueprints are designed with beginners in mind. They are plain and clear, with ample instructions. Use them, and you'll find the project goes much easier.

Most of this section of the book is divided into two parts, to explain two types of shades. Chapter 3, "Sizing the Shade," will help you decide which type you need for your windows. Once you have selected the type of shade you need, follow only the directions for that unit. If you read the directions for both units, you'll get confused. The two shades are very similar, but there are differences.

As you go through this section of the book, you'll soon realize that the most important step in making a shade that works well is sizing up your window and deciding where the shade should be mounted. For the majority of windows, the shade will be mounted in one of the two ways explained here. However, for special windows, you'll have to figure out special mounting methods on your own. The roller box and the fabric will stay the same, but the side closers will most likely need to be different. If you have a special installation, read all of section II and study the blueprints until you have a thorough understanding of how the shade works, then design a special mounting for your windows.

Actually, building the roller box and sewing the shade are quite easy; selecting the right location for the shade and roller box and measuring properly are the important and sometimes tricky parts.

Be sure you read the materials and tools sections before going any further. The materials section is especially important, as some of the materials we have available in our part of the country may not be available where you live. If you need to special-order anything, that should be done first. You most likely will have to mail-order the Mylar material, unless you choose to use "space blankets" from a local supplier.

The materials section tells you what we recommend, and why. From that you should be able to decide what types of substitutes will work for you. Use careful judgment, and if you don't know, take this book to your lumberyard or hardware store and ask. The folks who work in those stores are normally very knowledgeable and should help you out a great deal.

The heart and soul of this book are the data sheets. These sheets require only two measurements from your window; then with a few simple math calculations, you will have the exact size of every piece in the shade.

Chapter 3, "Sizing the Shade," goes over all the details of sizing the pieces in the shade. You must buy the fabric before you can finish sizing your shade, so as you work your way through this chapter you will also need to use chapter 4, "Sewing the Shade," to guide you in fabric selection.

Sewing the shade is very easy. All that is necessary is to cut the fabric to size and stitch edging material on three sides of it. Later, you have to make a hem on the fourth side. In some large installations, you may have to piece two sections of fabric together, which calls for some more advanced sewing, but still, it is only straight seam work.

Chapter 5, "Cutting the Pieces of the Roller Box," takes you through the cutting of each and every piece in the roller box. The design of the roller box is surprisingly simple. There are no joints or complex cuts. With one exception, every piece in the roller box is a simple rectangle. We give advice on what type of wood to buy. Follow that advice, and you'll end up with a better shade.

Chapter 6, "Assembling the Roller Box," goes through the details of assembling the roller box, and chapter 7, "Installing the Shade," covers the installation of the finished shade. Last, there is a short section on how to operate the shade.

We designed Rodale's Insulating Window Shade with two things in mind — energy conservation and you. You build it, and you save energy.

We feel this is the best insulating shade available. Because you build it yourself, it is also the least expensive for the amount of benefit you gain. Now it's time to start learning how to make an insulating shade and to start cutting down on those fuel bills.

TOOLS

SEWING MACHINE

A sewing machine is needed to bind the edges of the two pieces of fabric. See chapter 4 for sewing details.

SABER SAW

A saber saw is used to cut the pulley pieces and the rod holders. A coping saw will also work.

NAIL SET

Use a nail set to sink the heads of all nails.

WIRE CRIMPER

A wire crimper is used to fasten the pull cord butt splice.

WOOD CHISELS

Use ½-inch or ¾-inch wood chisels to cut recesses for the hinges on a window-mounted roller box. A smaller-size chisel is needed to make the opening for the slide bolt.

LEVEL

A level will help you ensure that the roller box is properly installed over the window.

MISCELLANEOUS TOOLS

Additional useful items are safety glasses, pencil, ruler, scissors, hammer, screwdriver, snap-setting equipment, and sandpaper or other abrasion devices such as Surform or Abrader.

COMBINATION SQUARE

The 45-degree angle on this square will be useful in making the rod holders, and the 90-degree angle will help square off edges on cut pieces.

DRILL (WITH ASSORTED BITS)

You will need a drill and assorted bits, primarily a 1/16-inch bit for pilot holes, and also some larger-size bits, a countersink, and a 1-inch (or the diameter size of your dowel rods) hole saw or spade bit for pulley and rod holder assembly.

CIRCULAR SAW (WITH RIP GUIDE)

A circular saw will make cutting long stretches of wood much easier, but if not available, a hand saw will do the job.

MATERIALS

WOOD

Either plywood or dimensional lumber can be used. We recommend plywood with an actual thickness of ¾ inch. What is sold as 1-inch dimensional lumber is actually ¾ inch thick. The amount of wood needed depends on your particular windows and the number of shades you wish to build. Complete the data sheet before buying your lumber. A small amount of ⅛- and ⅜-inch plywood is also needed. The bottom molding requires ¼-inch lattice.

WOOD FINISH

Paint, varnish, or stain can be used depending upon your preference.

VENEER TAPE

Veneer tape is needed only if plywood is used. It is available at hardware and building supply stores or lumberyards. There are two types available, self-adhesive or a type that requires glue; either will work.

DOWELS

Two are needed for each window shade. 1-inch-diameter dowels are best, but certain types of tool handles and round stock work well for wide windows.

FABRIC

The fabric should be quilted, preferably with a foam filling. Each shade requires two pieces. Read chapter 4 for fabric selection.

BIAS TAPE

This is used to finish off the edges of the fabric.

THREAD

The thread should match your fabric and bias tape.

CORD

Standard drapery cord and venetian blind cord are common terms for the type of cord you need.

ELECTRICAL BUTT SPLICE (UNINSULATED) FOR # 10 — 12 WIRE

Although this device is made for electrical wire, it will work perfectly for the cord on your shade.

SNAPS

There are two types of fabric snaps used. Chapter 7 has details on these snaps.

NAILS

Several types of nails are needed: 6d or 8d finishing nails — With the larger-size nails there is more chance of splitting the wood. 3d finishing nails — If you use a larger nail the point will stick through the wood on some construction steps. ½-inch brads — A very small number of brads will be needed when putting the pulleys together.

SCREWS

Standard flathead wood screws; several sizes are needed: 2½-inch # 10 flathead wood screws — These are needed to mount the finished roller box, unless you have a masonry wall or other special mounting condition. See chapter 7. 2-inch # 8 flathead wood screws — You will need these only if you are building a window-mounted shade. ¾-inch # 6 flathead wood screws — These screws are used for assembly of the bottom seal.

GLUE

Use any glue that is specified for wood. Epoxy glue may have to be used for imperfectly fitting pulleys.

SLIDE BOLTS

Two are needed for each shade. They are available in different styles; some are more decorative than others.

CORD TENSIONER

This device will hold your cord taut and keep it from getting tangled. It comes as a complete unit, and it is available at hardware, building supply, and drapery stores.

CAULKING

This is optional, but you may want to use it during installation of the shade.

HINGES

Two types of hinges are used: 1½-inch by 1½-inch open hinges — Two are needed for an average window shade; wider shades will need more. Clamping cabinet hinges — These are spring-loaded hinges and an average shade will need four or six of them. At least two are needed for each side of the shade.

RUBBER BANDS

Wide rubber bands are to be used on the pulleys to give added traction.

ALUMINIZED MYLAR

This is the plastic-based film with an extremely reflective aluminum coating. It is very important to the proper functioning of the shade, as it serves as a reflective barrier to stop radiant heat loss and to reduce infiltration. Aluminized Mylar normally will not be available from local stores. It can be ordered in 56-inch by 25-foot rolls from Kal-Wall Solar Components Division, Box 237, Manchester, NH 03105 (phone 603-668-8186), or in 4-foot by 25-foot rolls or 4-foot-wide pieces, cut to your length, from Solar Usage Now, 450 East Tiffin Street, Bascom, OH 44809 (phone 419-937-2226). The Mylar we have used is 5 mils thick and works very well. Drafting supply stores also handle aluminized Mylar. Some places sell 2-mil-thick Mylar which will work, but may not last as long. The company that makes aluminized Mylar is King Sealy, Metallized Products Division, Winchester, MA 01890 (phone 617-729-8300). Call and ask for a local distributor if you can't find a source. We have also used "space blankets," which work well for the winter mode but not as well for the summer mode, because they are colored on one side. "Space blankets" generally will cost more, since they are sold in set sizes, not rolls.

3 SIZING THE SHADE

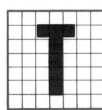he best thermal shade in the world will work poorly if made or installed improperly; correct installation and accurate measurements are crucial to making an effective shade. In this chapter, you will learn how to look at a window to determine where the shade should be mounted. Once this is done, a few simple measurements are all that are needed before starting to build your shade. Where the measurements are taken is the critical point.

This chapter has to be used in conjunction with chapter 4, "Sewing the Shade." You cannot calculate the size of the wooden pieces for the roller box until you have selected your fabric, nor can you buy fabric until you have figured out how much you will need. First read and use this chapter, then read chapter 4 and buy your fabric, then come back to this chapter to finish sizing your shade. The actual size of the roller box is determined, not only by the size of your window, but also by the thickness of the material you buy.

By the end of this chapter, you will know how to size a shade to custom-fit almost any window. Using charts 3-1 and 3-2 and the formulas on the data sheets, calculate the measurements for every piece.

Be sure to have additional blank copies of the data sheets photocopied so you'll be able to make shades for every window in the house.

Different window styles require different methods of mounting the shade. Rodale's

Insulating Window Shade is designed to be mounted in one of two places: on the wall directly above the window, as shown in photo 3-1; or, if you have deep windowsills, in the window cavity itself, as shown in photo 3-2.

First, we describe the procedure for wall-mounting a shade. If possible, you

should mount your shade that way. We also describe window-mounting, an alternative method for those cases where wall-mounting is not desirable.

The wall-mounted shade is our first choice, because it is the most efficient type of shade. Not only does it stop heat loss through

Photo 3-1 — A wall-mounted insulating window shade.

Photo 3-2 — A window-mounted insulating window shade.

the window, but it prevents infiltration of cold outside air into the house. A great deal of the air that enters your home through windows comes in through the trim work. The air finds its way through cracks on the outside of the house and ends up sneaking into the room through cracks around the window trim. A properly mounted shade will be caulked and will almost completely eliminate infiltration of cold outside air through the window trim. The side mounting boards are fastened on the wall outside the trim, enabling the shade to seal in the cold air, keeping it out of the room.

Although a wall-mounted shade is most efficient, for some windows it just won't work. For homes with very deep windowsills, having the shades mounted on the wall would mean inconveniences such as taking plants off the sills every night, making this type impractical. If you have a deep window, the critical measurement to determine if the shade can be mounted in the window cavity is the distance from the side of the window jamb to the start of the window channel. Illustration 3-1 shows where this measurement should be taken. You must have at least 1 inch of clearance at that point in order to mount a shade inside the window, or it won't fit.

The roller box for a window-mounted shade needs a minimum amount of space for the sides and pulleys at the top. However, the side closers and mounting boards of a window-mounted shade can be made larger to cover part of the window track, making the shade adaptable to a great many more win-

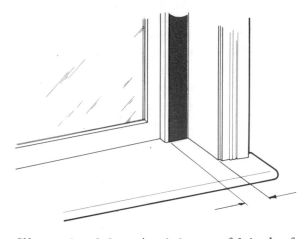

Illustration 3-1 — A minimum of 1 inch of clearance at the point shown is necessary to window-mount a shade.

dows, as you'll see in chapter 7, "Installing the Shade." If you have horizontally operated windows, or if you have deep windowsills with less than 1 inch of clearance between the jamb and the window, you will have to use wall-mounted shades.

You be the judge of where you want your shades if both mounting methods are possible; the differences in efficiency are minor. Before making your decision, read the sections on both wall-mounted and window-mounted shades to get a full understanding of how each works.

A WORD OF CAUTION: Mounting the shade on the window trim may seem inviting, but don't try it. Mounting it on the trim will reduce the infiltration seal, damage the trim, and most likely give you a shade that does not perform very well. Rodale's Insulating Window Shade is designed to mount either completely inside or completely outside the trim,

but not on it. *This is best; follow our advice.*

The rest of this chapter is in two parts. The first part tells how to size a shade that is to be mounted on the wall directly above the window. The second part tells how to size a shade for windows with deep windowsills where the shade mounts inside the window. Be careful that you use the right data sheet for each shade, as the measurements differ.

WALL-MOUNTED SHADE

To size a window for a wall-mounted shade, use illustration 3-2 and a wall-mounted shade data sheet (page 40). The illustration shows where to take measurements A, B, and C. Measure carefully, and double-check. A mistake at this time will make the entire shade the wrong size.

All measurements should be taken and recorded in inches. When taking measurements, round the exact distance downward to the nearest ⅛ inch.

Measurement A is taken from the outside of the trim on one side to the outside of the trim on the other side of the window. Take this reading in several places to compensate for any small abnormalities in the trim. If your windows have larger pieces of trim at the corners, as shown in illustration 3-11 (page 36), don't measure these; doing so would give you a shade too large for the window. Measure along the long, even run of trim, as shown in illustration 3-2.

Measurement B is taken from the top

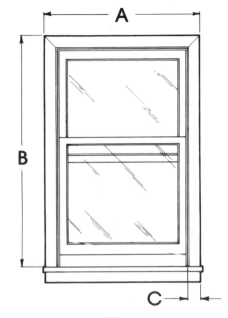

Illustration 3-2 — Measurement points for a wall-mounted shade.

edge of the top molding to the top of the sill. Take this reading on both sides of the window. If the readings are only slightly different, use the larger of the two measurements. You can compensate for any unevenness when installing the roller box or when cutting the fabric to its finished length. If the difference between the two measurements is greater than ½ inch, you have a problem window. Using a level, see if the window frame is uneven, if the trim is off, or if you just measured wrong. If the window is out of line, use the largest measurement for now, and remember that you will have to adjust for the unevenness later.

Measurement C is the width of the trim.

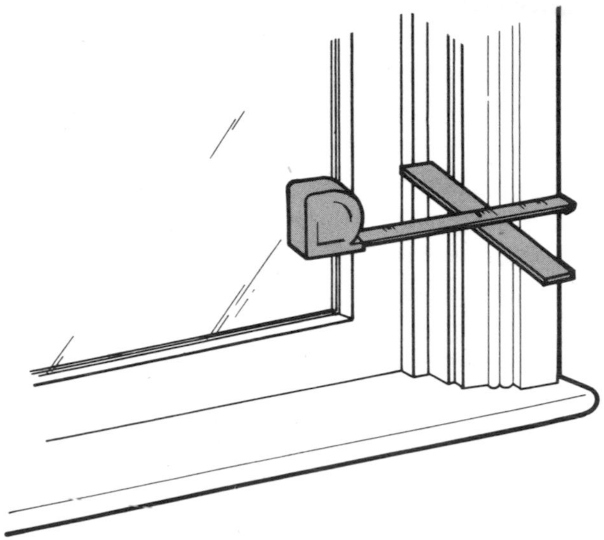

Illustration 3-3 — Take measurement C with a ruler and a tape measure.

This measurement determines whether or not you have enough surface on which to seal the shade. A minimum of 2 inches is recommended. To take this measurement, hold a straightedge against the inside edge of the trim, and measure from the outside of the trim to the edge of the straightedge, as shown in illustration 3-3.

If you have a window with no trim, you have two choices. You can either install trim around the window frame and then proceed as already explained, or you can expand the size of the side mounting boards when installing the shade, and seal the fabric against them. In the latter case, take measurement A from one inside edge of the window to the other (as shown for the window-mounted shade in illustration 3-11 on page 36), and add a total of 4 inches to the measurement before recording it on the data sheet. This will allow for an additional 2 inches of mounting board to be added to each side for the fabric to seal against.

Record all three measurements at the top of the data sheet. Just to be safe, go back and remeasure all three points, especially measurements A and B. Double-check your figures. If one of these measurements is wrong, the entire shade will be off. Now is the time to catch any errors in measurement.

With measurements A, B, and C recorded on the data sheet, you are ready to begin figuring out the sizes of the pieces in the shade. The first step is to calculate the amount of fabric you will need.

AMOUNT OF FABRIC
To arrive at the finished width of the fabric, simply subtract ¼ inch from measurement A. Remember, this is the finished width of the fabric. Allow some additional fabric for waste as explained in chapter 4, "Sewing the Shade."

To estimate the length of the fabric, add 12 inches to measurement B. This will give you the estimated length of the fabric. The actual length will not be determined until the roller box is installed and the fabric hung.

With these two fabric measurements recorded, read about fabric selection in chapter 4, "Sewing the Shade." Then go to the fabric store and buy your fabric. Once you have selected your fabric, return to this chap-

		R-Value	Percent of Heat Saved vs. Single-Glazed Window	Fabric Classification Number

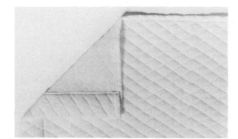

Quilted taffeta with mesh backing — One side is a smooth, glossy, transparent mesh, and the other side is a closer weave. It has about $\frac{1}{16}$ inch of synthetic fill and is extremely lightweight.

| 4.08 | 77% | # 1 |

Polyester with lining — This has two sides of a shiny, lightweight synthetic, finely woven, with a vinyl feel. It is vertically quilted, with about $\frac{3}{16}$ inch of a loosely matted, fibrous synthetic fill.

| 4.98 | 81% | # 2 |

Cotton with lining — This has a coarse, heavy-duty weave on one side and a fine, linen-like white backing with about $\frac{3}{16}$ inch of spun cotton, loosely matted, as fill.

| 5.04 | 81% | # 3 |

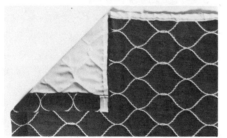

Place mat — This has a linen-like cotton material on both sides with $\frac{1}{4}$ inch of foam quilted between them. It has a more stable construction than the other fabrics, due to the foam filling.

| 4.83 | 80% | # 4 |

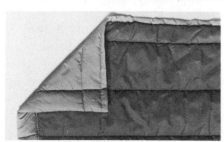

Ski jacket (fiberfill) — This has two sides of finely woven, lightweight nylon. It is quilted in a wide vertical pattern with $\frac{1}{2}$ inch of synthetic fiber.

| 8.01 | 88% | # 5 |

Chart 3-1 — Fabric Classification

ter and continue the sizing process. You cannot properly size the shade until you have fabric.

After you have bought your fabric, look at chart 3-1 to find the thickness classification number of your fabric. The chart shows five different thicknesses of material, ranging from very thin taffeta to thick ski jacket material. Decide which fabric type is closest to what you have bought, and note its classification. If your fabric does not fall exactly into one of the classifications, use the next larger listing. When in doubt, go up a classification size. The chart lists the thinnest fabric as # 1 and the thickest as # 5.

Once you have your fabric's thickness classification noted, go to chart 3-2. Find the column on top of the chart that matches the classification number you found in chart 3-1. Then, look at the left-hand column, and find the listing for the ROUGH fabric length of your window, from your data sheet. Read across until you reach the fabric classification column for your fabric. The number in the corresponding box is the center-to-center distance, in inches, for your shade.

The center-to-center distance equals the diameter of a roll of fabric for your length of window, including a 1-inch dowel rod. Record this figure on your data sheet. This is a very important figure, as it determines many of the dimensions of the roller box. If the distance you calculate is too small, there will be too much pressure on the rollers when you roll the shade up, and the shade will not work smoothly. If you overestimate your fabric's thickness, the shade will work fine, but the

roller box will be slightly larger than necessary. It is far better to make the roller box larger than to have a shade that does not roll well. We have allowed for some extra play in the figures in chart 3-2. Don't add any additional space on your own.

Once you have the center-to-center distance calculated and recorded on your data sheet, you can begin to size the wooden pieces of the shade. Start at the top of the data sheet and work your way down. You will

arrive at two figures for each piece in the shade with the exception of the dowels and pulleys. The drawings of each piece will help you figure out how to calculate each measurement. Blueprint sheet 1 will also help you clarify what piece you are working on, and how to size it properly.

All the measurements and calculations given are for a roller box made with ¾-inch stock and 1-inch-diameter dowels. If you want to use material of another thickness, we

Chart 3-2 — Center-to-Center Distance in Inches

Rough Fabric Length in Inches	Fabric Classification Number				
	# 1	# 2	# 3	# 4	# 5
Up to 24	3	3	3	3	3
25 – 30	3	3	3	3	$3\frac{1}{4}$
31 – 36	3	3	3	$3\frac{1}{4}$	$3\frac{1}{4}$
37 – 42	3	3	3	$3\frac{1}{2}$	4
43 – 48	3	$3\frac{1}{4}$	$3\frac{1}{4}$	$3\frac{3}{4}$	$4\frac{1}{4}$
49 – 54	$3\frac{1}{4}$	$3\frac{1}{4}$	$3\frac{1}{2}$	4	$4\frac{1}{4}$
55 – 60	$3\frac{1}{4}$	$3\frac{1}{2}$	$3\frac{1}{2}$	4	$4\frac{3}{4}$
61 – 66	$3\frac{1}{2}$	$3\frac{3}{4}$	$3\frac{3}{4}$	$4\frac{1}{4}$	5
67 – 72	$3\frac{1}{2}$	4	4	$4\frac{1}{2}$	$5\frac{1}{4}$
73 – 78	$3\frac{3}{4}$	$4\frac{1}{4}$	$4\frac{1}{4}$	$4\frac{3}{4}$	$5\frac{1}{2}$
79 – 84	4	$4\frac{1}{2}$	$4\frac{1}{2}$	$4\frac{3}{4}$	$5\frac{3}{4}$
85 – 90	$4\frac{1}{4}$	$4\frac{3}{4}$	$4\frac{3}{4}$	5	6

give (in italic type) an explanation of how the calculations for each piece were reached. However, we recommend that ¾-inch stock be used so that the simple calculations on the data sheet can be used.

PIECE # 1 — BACK
The length of the back equals measurement A plus 3 inches. The height equals two times the center-to-center distance.

If you are using wood of a different thickness, the length measurement is calculated by adding the combined thicknesses of the two rod holders (piece # 4) to measurement A, plus ⅝ inch for the pulley (piece # 8), an additional ⅝ inch for space on the other end to keep the box centered over the window, and a total of ¼ inch to allow ⅛ inch of clearance at each end. The sum of these will give you the proper length of the back. The height of the back is not influenced by the thickness of the wood you use.

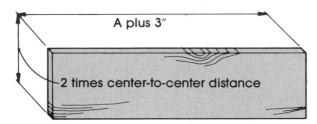

Illustration 3-4 — Piece # 1 — back.

PIECE # 2 — TOP
The length of the top is calculated the same as the back, measurement A plus 3 inches. The width of the top is calculated by adding 1½ inches to the center-to-center distance.

If you use wood of a different thickness, the length of the top is figured the same way as the length of the back (piece # 1). The width is found by taking the center-to-center distance and adding the thickness of the back (piece # 1) plus an additional ¾ inch, to allow ⅜ inch of clearance in both the front and the back.

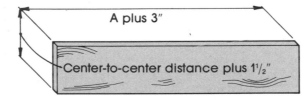

Illustration 3-5 — Piece # 2 — top.

PIECE # 3 — SIDES
You need two sides for the roller box. The height of the sides is calculated by adding 2¼ inches to two times the center-to-center distance. Note: That is TWO times the center-to-center distance, plus 2¼ inches. The width

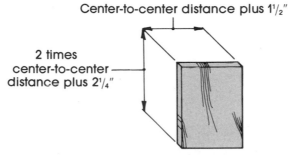

Illustration 3-6 — Piece # 3 — side.

of the sides is found by adding 1½ inches to the center-to-center distance.

If you are using wood of a different thickness, the height is calculated by first taking the center-to-center distance and doubling it. To that figure, add ¾ inch for a bottom extension, and then add the thicknesses of both the top (piece # 2) and the bottom (piece # 6). The width of the sides is calculated by taking the center-to-center distance and adding the thickness of the back (piece # 1) plus ¾ inch to allow ⅜ inch of clearance in both the front and the back.

PIECE # 4 — ROD HOLDERS
The actual rod holders are in two pieces, referred to as pieces A and B. Piece A is mounted to the roller box, and piece B is mounted to the hinged front of the roller box. At this point, all we will do is size the piece of wood from which the rod holders will be cut. The height is found by taking the center-to-center distance and multiplying by two. The width is

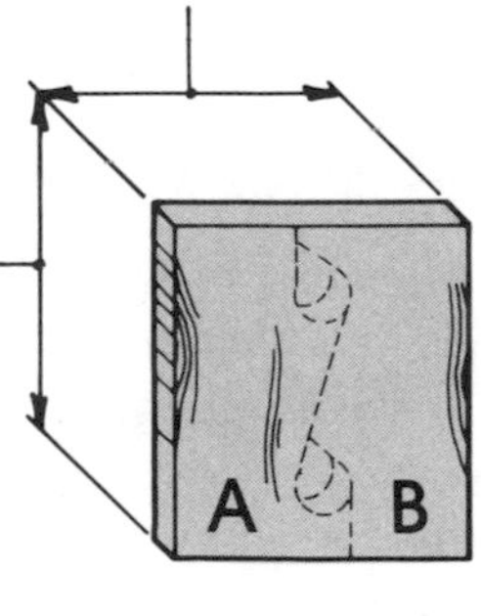

Illustration 3-7 — Piece # 4 — rod holder.

found by taking the center-to-center distance and adding ¾ inch.

The height and width of the rod holders stays the same, regardless of wood thickness.

PIECE # 5 — FRONT

The length equals measurement A plus 4½ inches. The height equals two times the center-to-center distance plus 2¼ inches.

If you are using wood of a thickness other than ¾ inch, calculate the size of the front by taking measurement A and adding the thicknesses of the two sides (piece # 3) and the two rod holders (piece # 4). With those four numbers added to measurement A, add 1½ inch to allow for ⅝ inch for the pulley on one side and ⅝ inch of space on the other side and ¼ inch for clearance, to get the total length of the piece. To find the height, start with two times the center-to-center distance. To that figure, add the thickness of the top (piece # 2), the thickness of the bottom (piece # 6), and an extra ¾ inch for a bottom extension.

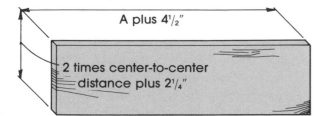

Illustration 3-8 — Piece # 5 — front.

PIECE # 6 — BOTTOM

The length of the bottom is found by adding 3 inches to measurement A. The width is the center-to-center distance.

If you are using wood of a thickness other than ¾ inch, the length of the bottom is calculated by the same method used for finding the length of the back (piece # 1). To measurement A, add the thicknesses of the two rod holders (piece # 4) plus 1¼ inches to allow for a pulley at one end and space at the other end. Then add ¼ inch to give clearance at each end. The width is found by taking the center-to-center distance. The width measurement is not affected by the thickness of your wood.

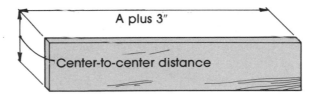

Illustration 3-9 — Piece # 6 — bottom.

PIECE # 7 — DOWEL RODS

Each roller box uses two dowel rods. These should be 1 inch in diameter. Their length is determined by adding 2⅞ inches to measurement A.

Most hardware stores carry dowels of various diameters, but only in 3-foot lengths. Round stock comes in both 1-inch and 1¼-inch diameter; either size will work for a roller box. If you can't get the proper size dowel or round stock, look for straight tool handles. Tool handles often come in 6-foot lengths and are normally the necessary 1 inch in diameter. Another source of dowel rods in longer lengths is the drapery department of hardware and home stores.

If you are using wood of a thickness other than ¾ inch, calculate the length of the dowels the same way you calculated the length of the back (piece # 1), but subtract ⅛ inch from the calculated length.

PIECE # 8 — PULLEYS

There are two pulleys in each roller box. Each pulley is made from three circles of wood. In chapter 5, "Cutting the Pieces of the Roller Box," we will describe what thickness material to use and how to assemble the pieces. For now, all you have to do is determine the diameter of the circles. The diameter of the sides of the pulleys is calculated by

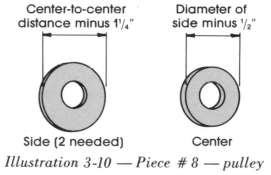

Illustration 3-10 — Piece # 8 — pulley parts.

taking the center-to-center distance and subtracting 1¼ inches. If this figure is less than 2 inches, make the pulleys 2 inches. A pulley smaller than 2 inches will not work.

The diameter of the center piece of the pulley is found by subtracting ½ inch from the diameter of the sides of the pulleys.

SIZING THE REMAINING PIECES

The sizing of the side closers, mounting boards, and bottom pieces will be done after the shade is installed. These will be covered in chapter 7, "Installing the Shade." For now, you are done sizing the pieces of your shade. Proceed to chapter 4, "Sewing the Shade," and begin sewing the fabric for your shade.

WINDOW-MOUNTED SHADE

This type of shade is designed for mounting directly in the window opening. It is mainly for windows with deep sills, but some people may choose to install it in relatively shallow window frames.

The main difference between this type of shade and a wall-mounted shade is that this unit has no side pieces. The rod holders themselves form the sides. For aesthetic reasons, this is not too attractive for a wall-mounted installation. If you have windowsills deep enough to hide the sides of the roller box, this is a good type of installation.

To size a window-mounted shade, first look at illustration 3-11 to see where to take the measurements. Each measurement should be recorded at the top of the window-mounted shade data sheet (these differ from the wall-mounted shade data sheets) on page 40.

The first measurement to be taken is measurement A. This is the distance from the inside of one side of the window jamb to the opposite side of the jamb, the area that the roller box will actually fill. Measure this at the top of the window. Be sure to measure at the actual window, not at the front of the window frame opening. Many old-style windows have slightly angled jambs, and a measurement taken at the opening will be larger than one taken right at the window where the roller box will be mounted. A roller box made using the larger measurement would not fit into the window jamb. To put it simply, measure exactly where the box will be installed.

Measurement B is the height of the window opening. It is measured from the top of the window opening to the sill. Measure on both sides of the window opening to be sure the window is fairly square.

Measurement C is the amount of window frame available on which to install the mounting boards and side closers. If there is less than 1 inch of space between the window jamb and the actual window track, you will not be able to mount the shade in the window, and it will have to go on the wall. Illustration 3-1 (page 30) shows where this measurement should be taken. Be sure you understand what you are to measure before taking measurement C.

Measurement D is the depth of the window opening. If this measurement is less than the width of the sides of the roller box, part of the box will protrude from the window opening. If this is the case, you may want to switch to a wall-mounted unit for aesthetic reasons. For now, record measurement D, and after sizing the pieces of the box, come

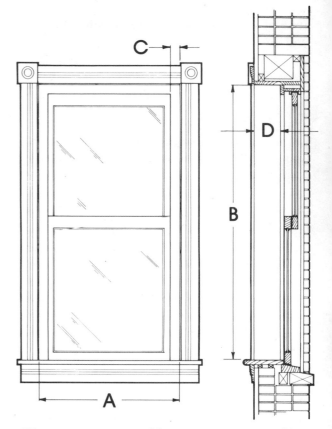

Illustration 3-11 — Measurement points for a window-mounted shade.

back to it to see if part of the box will show outside of the window frame.

In some windows, the trim along the wall overhangs the window opening, making measurement A smaller at the window opening along the wall than at the window itself, as shown in illustration 3-12. If this is the case on your window, measure the distance from the inside edges of this trim, as shown in illustration 3-12. You will have to calculate whether the front of the box will have enough room to open when installed and whether you will even be able to fit the roller box past the trim to install it. If the front won't have enough clearance, or if the box won't fit in the opening, you'll have to use a wall-mounted unit.

Record all measurements in inches on the window-mounted shade data sheet. When taking a measurement, round the exact figure downward to the nearest ⅛ inch. With the measurements written on the form, go back and remeasure each area to be sure no mistakes were made.

With all four measurements recorded, you are ready to begin sizing the shade. The first step in sizing a window-mounted shade is to calculate the fabric size. To find the finished width of the fabric, take measurement A and subtract 3 inches. Remember that this is the finished width; you will need a little extra fabric for waste.

To find the length of the fabric, take measurement B and add 6 inches. This is only a rough measurement; the fabric will be cut to its final length after the roller box is installed.

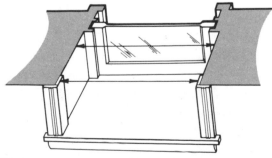

Illustration 3-12 — Trim often overhangs the window opening; measure carefully.

Record the fabric measurements on the data sheet. Turn to chapter 4, "Sewing the Shade," and read about fabric selection. Once you understand what types of fabrics are best, go to a fabric store and buy your fabric. You must have the fabric before continuing to size the shade.

FABRIC CLASSIFICATION

When you have bought your fabric, go to chart 3-1 (page 32) to find what thickness your fabric should be classified. The chart shows five different thicknesses of material, ranging from the thinnest taffeta we have used to the thickest ski jacket fabric we've used. Decide which fabric type is closest to what you have purchased, and note what classification number it falls under. If your fabric does not come directly under one classification, use the next larger fabric listing. When in doubt, go up a classification size. The chart classifies the thinnest fabric as # 1 and the thickest as # 5.

Once you know your fabric's classification number, go to chart 3-2 (page 33). Find the classification number of your fabric across the top of the chart. Then, in the left-hand column, find the listing for the rough fabric length of your window, as recorded on your data sheet. Read across the listing until you reach the fabric classification column for your fabric. The number that appears here is the center-to-center distance for your shade.

The center-to-center distance equals the diameter of a roll of fabric for your length of window, including a 1-inch dowel. Record

the center-to-center distance on your data sheet. This is a very important figure, as it determines many of the dimensions of the roller box. Go back over the calculation process one more time to be sure you have made no errors.

If the center-to-center distance you calculate is too small, there will be too much pressure on the fabric when it rolls up, and the shade will not work smoothly. If you overestimate your fabric's thickness, the shade will work just fine, but the roller box will be slightly larger than necessary. It is far better to err on the side of a larger roller box than to have a shade that does not roll well. We have allowed for some extra play in the figures in chart 3-2; don't add any additional space on your own.

With the center-to-center distance calculated and recorded on your data sheet, you are ready to begin sizing the wooden pieces of the shade. Start at the top of the data sheet and work your way down. You will calculate two figures for each piece in the shade, with the exception of the dowel rods and pulleys. If you have any questions or misunderstandings of the dimensions of a piece, refer to blueprint sheet 2. The drawings of each piece should help you figure out how to calculate each measurement.

All the measurements and calculations on the data sheet are for a roller box made with ¾-inch stock and 1-inch-diameter dowels. If you want to use material of another thickness, we give an explanation of how the calculations for each piece were reached. These instructions are in italic type; disregard them if you are using the recommended ¾-inch material.

The data sheet sizing system is one of the unique features of Rodale's Insulating Window Shade, and it allows you to custom-make a shade for each particular window. Take your time and carefully complete your data sheet.

PIECE # 1 — BACK
The first piece to size is the back. The length is calculated by taking measurement A and subtracting 1⅝ inches. The height is found by doubling the center-to-center distance and adding 1½ inches.

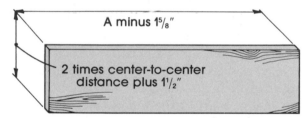

Illustration 3-13 — Piece # 1 — back.

If you are using material of a thickness other than ¾ inch, calculate the length of the back by subtracting the thickness of each of the two rod holders (piece # 3) from measurement A. Subtract an additional ⅛ inch for clearance. The height of the back is found by doubling the center-to-center distance, adding the thickness of the bottom (piece # 5), and adding an additional ¾ inch for clearance.

PIECE # 2 — TOP
The size of the top is found by taking measurement A and subtracting ⅛ inch for the length. Add 2¼ inches to the center-to-center distance to get the width.

If you are using wood of a thickness other than ¾ inch, calculate the length of the top by subtracting ⅛ inch from measurement A, regardless of what thickness of wood you use. To find the width, take the center-to-center distance and add ¾ inch for clearance inside the box, plus the combined thicknesses of the back (piece # 1) and the front (piece # 4).

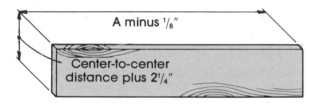

Illustration 3-14 — Piece # 2 — top.

PIECE # 3 — ROD HOLDERS
The rod holders are actually four pieces of the roller box. They are used in pairs and referred to as pieces A and B. At this time we will size the two pieces from which the four pieces are cut.

To find the height, take two times the center-to-center distance, and add 1½ inches. To find the width, take the center-to-center distance, and add 1½ inches.

If you are using material of a thickness other than ¾ inch, calculate the height of the rod holders the same way you did the height of the back (piece # 1). Take two times the center-to-center distance, add ¾ inch for the

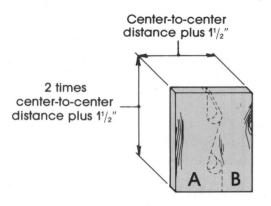

Illustration 3-15 — Piece # 3 — rod holders.

bottom extension, and add the thickness of the bottom (piece # 5). To figure the width, take the center-to-center distance, add ¾ inch for inside clearance, and add the thickness of the back (piece # 1).

PIECE # 4 — FRONT
To calculate the length of the front, subtract ⅛ inch from measurement A. For the height, take two times the center-to-center distance, and add 1½ inches.

If you are using material other than ¾ inch thick, calculate the length as explained above and on the data sheet. The thickness of the wood does not affect this measurement.

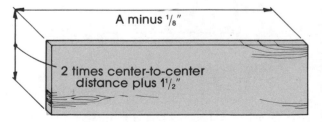

Illustration 3-16 — Piece # 4 — front.

Find the height by taking two times the center-to-center distance, adding ¾ inch for a bottom extension, and adding the thickness of the bottom (piece #5).

PIECE # 5 — BOTTOM
The length of the bottom is calculated by taking measurement A and subtracting 1⅝ inches. The width is the center-to-center distance.

If you are using material of a thickness other than ¾ inch, calculate the length by subtracting ⅛ inch for clearance from measurement A, and then subtracting the combined thickness of the two rod holders (piece #3). The width of the bottom is found as explained above, regardless of the thickness of wood used.

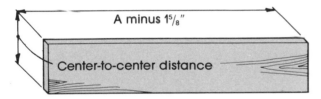

Illustration 3-17 — Piece #5 — bottom.

PIECE # 6 — DOWEL RODS
Each roller uses two dowel rods. These should be 1 inch in diameter. Their length is determined by subtracting ⅛ inch from measurement A.

Most hardware stores carry dowels of various diameters, but only in 3-foot lengths. Round stock is available in both 1-inch and 1¼-inch diameter in varying lengths, often up to 7 feet. The larger-size dowel will work

with no changes needed in the sizing procedure of your shade. If you can't find a dowel or round stock of the proper length, look for tool handles, as they normally come in 1-inch diameter sizes. Another source of dowel rods in longer lengths is the drapery department of hardware and home stores. As a last resort, go to a custom drapery shop for longer dowel rods.

The length calculation for the dowel rods does not change, regardless of what thickness of wood you use for your roller box.

PIECE # 7 — PULLEYS
There are two pulleys in each roller box. Each pulley is made from three circles of wood. In chapter 5, "Cutting the Pieces of the Roller Box," we describe what thickness of material to use, and in chapter 6, "Assembling the Roller Box," how to put the pieces together. For now, just determine the diameter of the circles.

The diameter of the sides of the pulleys is calculated by taking the center-to-center distance and subtracting 1¼ inches. If the

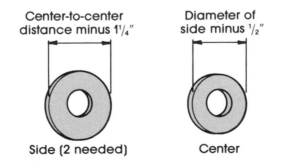

Illustration 3-18 — Piece #7 — pulleys.

figure you arrive at is less than 2 inches, make the pulleys 2 inches in diameter. A pulley smaller than 2 inches will not work properly.

The diameter of the center of the pulley is found by subtracting ½ inch from the diameter of the pulley sides.

The diameter of the pulleys does not change, regardless of the thickness of wood used for the roller box.

SIZING THE REMAINING PIECES
The sizing procedure for the side closers, mounting boards, and bottom pieces will be covered in chapter 7, "Installing the Shade." These pieces are best calculated after the roller box has been installed.

At this point you should have a completely filled out data sheet. You are now ready to go to the next chapter and sew your shade. Then you will use the data sheet to cut out the wood for your shade and assemble the pieces. Last, you will install the shade. For now, let's go on to sewing the shade.

Wall-Mounted Shade Data Sheet

Window Location _____

FINISHED FABRIC WIDTH:

Measurement A _____ Measurement B _____ Measurement C _____

Measurement A MINUS $1/4$" = _____ Finished Fabric Width

ROUGH FABRIC LENGTH:

Measurement B PLUS 12" = _____ Rough Fabric Length

FABRIC CLASSIFICATION NUMBER = _____

CENTER-TO-CENTER DISTANCE
(from chart 3-2) = _____

PIECE #	NAME	QUANTITY NEEDED	CALCULATION	DIMENSION	
1	Back	1	LENGTH = A PLUS 3" WIDTH = 2 TIMES Center-to-Center Distance	LENGTH WIDTH	___ ___
2	Top	1	LENGTH = A PLUS 3" WIDTH = Center-to-Center Distance PLUS $1^1/2$"	LENGTH WIDTH	___ ___
3	Sides	2	HEIGHT = 2 TIMES Center-to-Center Distance PLUS $2^1/4$" WIDTH = Center-to-Center Distance PLUS $1^1/2$"	HEIGHT WIDTH	___ ___
4	Rod Holders	2	HEIGHT = 2 TIMES Center-to-Center Distance WIDTH = Center-to-Center Distance PLUS $3/4$"	HEIGHT WIDTH	___ ___
5	Front	1	LENGTH = A PLUS $4^1/2$" HEIGHT = 2 TIMES Center-to-Center Distance PLUS $2^1/4$"	LENGTH HEIGHT	___ ___
6	Bottom	1	LENGTH = A PLUS 3" WIDTH = Center-to-Center Distance	LENGTH WIDTH	___ ___
7	Dowel Rods	2	LENGTH = A PLUS $2^7/8$"	LENGTH	___
8	Pulleys	2	DIAMETER OF SIDES = Center-to-Center Distance MINUS $1^1/4$" (not to be smaller than 2") DIAMETER OF CENTER = Diameter of Sides MINUS $1/2$"	SIDE DIAMETER CENTER DIAMETER	___ ___

Wall-Mounted Shade Data Sheet

Window Location _____

Measurement A _____   Measurement B _____   Measurement C _____

FINISHED FABRIC WIDTH:

Measurement A MINUS ¹/₄" = _____ Finished Fabric Width

ROUGH FABRIC LENGTH:

Measurement B PLUS 12" = _____ Rough Fabric Length

FABRIC CLASSIFICATION NUMBER = _____

CENTER-TO-CENTER DISTANCE
(from chart 3-2) = _____

PIECE #	NAME	QUANTITY NEEDED	CALCULATION	DIMENSION
1	Back	1	LENGTH = A PLUS 3" WIDTH = 2 TIMES Center-to-Center Distance	LENGTH ___ WIDTH ___
2	Top	1	LENGTH = A PLUS 3" WIDTH = Center-to-Center Distance PLUS 1¹/₂"	LENGTH ___ WIDTH ___
3	Sides	2	HEIGHT = 2 TIMES Center-to-Center Distance PLUS 2¹/₄" WIDTH = Center-to-Center Distance PLUS 1¹/₂"	HEIGHT ___ WIDTH ___
4	Rod Holders	2	HEIGHT = 2 TIMES Center-to-Center Distance WIDTH = Center-to-Center Distance PLUS ³/₄"	HEIGHT ___ WIDTH ___
5	Front	1	LENGTH = A PLUS 4¹/₂" HEIGHT = 2 TIMES Center-to-Center Distance PLUS 2¹/₄"	LENGTH ___ HEIGHT ___
6	Bottom	1	LENGTH = A PLUS 3" WIDTH = Center-to-Center Distance	LENGTH ___ WIDTH ___
7	Dowel Rods	2	LENGTH = A PLUS 2⁷/₈"	LENGTH ___
8	Pulleys	2	DIAMETER OF SIDES = Center-to-Center Distance MINUS 1¹/₄" (not to be smaller than 2") DIAMETER OF CENTER = Diameter of Sides MINUS ¹/₂"	SIDE DIAMETER ___ CENTER DIAMETER ___

Window-Mounted Shade Data Sheet

Window Location _____

Measurement A _____   Measurement B _____   Measurement C _____
                         Measurement D _____

FINISHED FABRIC WIDTH:
Measurement A MINUS 3" = _____ Finished Fabric Width

ROUGH FABRIC LENGTH:
Measurement B PLUS 6" = _____ Rough Fabric Length

FABRIC CLASSIFICATION NUMBER = _____

CENTER-TO-CENTER DISTANCE
(from chart 3-2) = _____

PIECE #	NAME	QUANTITY NEEDED	CALCULATION	DIMENSION
1	Back	1	LENGTH = A MINUS $1\frac{5}{8}$" WIDTH = 2 TIMES Center-to-Center Distance PLUS $1\frac{1}{2}$"	LENGTH _____ WIDTH _____
2	Top	1	LENGTH = A MINUS $\frac{1}{8}$" WIDTH = Center-to-Center Distance PLUS $2\frac{1}{4}$"	LENGTH _____ WIDTH _____
3	Rod Holders	2	HEIGHT = 2 TIMES Center-to-Center Distance Plus $1\frac{1}{2}$" WIDTH = Center-to-Center Distance PLUS $1\frac{1}{2}$"	HEIGHT _____ WIDTH _____
4	Front	1	LENGTH = A MINUS $\frac{1}{8}$" HEIGHT = 2 TIMES Center-to-Center Distance PLUS $1\frac{1}{2}$"	LENGTH _____ HEIGHT _____
5	Bottom	1	LENGTH = A MINUS $1\frac{5}{8}$" WIDTH = Center-to-Center Distance	LENGTH _____ WIDTH _____
6	Dowel Rods	2	LENGTH = A MINUS $1\frac{1}{8}$"	LENGTH _____
7	Pulleys	2	DIAMETER OF SIDES = Center-to-Center Distance MINUS $1\frac{1}{4}$" (not to be smaller than 2") DIAMETER OF CENTER = Diameter of Sides MINUS $\frac{1}{2}$"	SIDE DIAMETER _____ CENTER DIAMETER _____

Window-Mounted Shade Data Sheet

Window Location _____

Measurement A _____   Measurement B _____   Measurement C _____
                        Measurement D _____

FINISHED FABRIC WIDTH:
Measurement A MINUS 3" = _____ Finished Fabric Width

ROUGH FABRIC LENGTH:
Measurement B PLUS 6" = _____ Rough Fabric Length

FABRIC CLASSIFICATION NUMBER = _____

CENTER-TO-CENTER DISTANCE
(from chart 3-2) = _____

PIECE #	NAME	QUANTITY NEEDED	CALCULATION	DIMENSION
1	Back	1	LENGTH = A MINUS 1⅝" WIDTH = 2 TIMES Center-to-Center Distance PLUS 1½"	LENGTH __ __ WIDTH __ __
2	Top	1	LENGTH = A MINUS ⅛" WIDTH = Center-to-Center Distance PLUS 2¼"	LENGTH __ __ WIDTH __ __
3	Rod Holders	2	HEIGHT = 2 TIMES Center-to-Center Distance Plus 1½" WIDTH = Center-to-Center Distance PLUS 1½"	HEIGHT __ WIDTH __
4	Front	1	LENGTH = A MINUS ⅛" HEIGHT = 2 TIMES Center-to-Center Distance PLUS 1½"	LENGTH __ __ HEIGHT __ __
5	Bottom	1	LENGTH = A MINUS 1⅝" WIDTH = Center-to-Center Distance	LENGTH __ __ WIDTH __ __
6	Dowel Rods	2	LENGTH = A MINUS ⅛"	LENGTH __
7	Pulleys	2	DIAMETER OF SIDES = Center-to-Center Distance MINUS 1¼" (not to be smaller than 2") DIAMETER OF CENTER = Diameter of Sides MINUS ½"	SIDE DIAMETER __ CENTER DIAMETER __

In this chapter we will first discuss the types of fabrics that are available, their pros and cons, and how they can affect the thermal performance of the shade. Then we will describe the sewing of the shade.

The fabric is the part of your shade you will be looking at all the time, so be sure you pick a fabric you like, and sew it properly. Sewing the fabric is actually quite simple; the type of, and the design on, the fabric is more important than how you sew it.

For maximum thermal value, the shade needs a multilayered fabric. The more layers of fabric, the higher the insulating value of the finished shade. However, there must be some way of keeping the different layers together as a unit. For that reason, we have chosen to use two separate layers of quilted material for the shade.

The two layers of quilted material are sandwiched around a single layer of Mylar. As illustration 4-1 shows, that gives a total of seven layers in the shade. The Mylar is not sewn to the fabric; it is attached at the top by snaps. The unsewn Mylar thus provides a perfect, unbroken vapor and infiltration barrier. Thanks to the Mylar barrier, it does not matter that the quilted fabrics are sewn; they do not serve as an infiltration barrier, only an insulating layer. We will discuss how to cut and fasten the Mylar in chapter 7, "Installing the Shade." At this time, we will only be concerned with the two fabric layers.

Most fabric stores have a large selection of quilted material available. These fabrics are reasonably priced and will work well for an insulating shade. Generally, fabric stores will stock up on quilted materials in the fall.

Be aware that fabric stores do not stock huge inventories of quilted fabrics. If you are going to be making a number of shades, go to your fabric store early and let them know what type of fabric you will be needing and how much. They can order the material you need, in the amount you need. Otherwise, you'll find yourself using many different types of fabrics and running short of fabric. Be warned that you should preorder your fabric if you plan to make any number of shades and want the same fabric for several or all of them. Normally, quilted materials are sold in small quantities. Insulating shades require relatively large amounts of fabric, up to four yards a window. Have enough fabric preordered for the number of shades you plan to make.

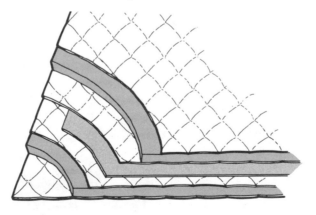

Illustration 4-1 — A Mylar sheet sandwiched between two pieces of quilted fabric gives a total of seven layers.

Most quilted materials consist of two layers of fabric, with a middle layer of some type of padding. The outer layers may be patterned on one or both sides, solid on one or both sides, or a combination.

Although they may look as though they would work well, avoid fabrics with loose insulating fill. Many ski jacket materials use down or a synthetic loose filling. These have very good insulating qualities, but must be worn regularly to keep the insulating material from bunching up. In an insulating shade the loose insulating material would quickly accumulate at the bottom of the shade, robbing the top part of the shade of insulating value. If you select a fabric with a good, close quilting pattern, this would not be as much of a problem. However, if you pick a fabric with a vertical (tube-type) quilting pattern, this would be a concern. Use a fabric with an insulating material that will not move around and change position.

The actual type of fabric used to make the quilted material does not matter too much. We have used fabrics made with 100 percent cotton, milled blends, and 100 percent artificial fabric, and all have worked well. As long as the fabric can be sewn easily, and can be washed, it will work for your shade. Some of the man-made fabrics are quite slick and may not feed evenly through your sewing machine. Be sure you get a fabric your machine can handle.

When looking at quilted materials be sure to look at both sides; a plain backing on an otherwise gaudy print may be just what you want for your shade. Only one side of the fabric shows. If you find a print you like, but really hate the other side, remember the ugly side can be put on the inside of the shade, and no one will ever see it. Some of the less expensive quilted materials have good fabric on one side, and only a lightweight mesh to hold the quilting in place on the other side. These fabrics work well, as the two unfinished sides can face each other while the two finished sides face out.

You also may want to find a print with a plain white backing, use the white side for the shade side that faces outside, and use a pattern you like for the side that faces into the room. *The thing to remember is that both layers of the shade must be of fabric of the same thickness.*

You can mix the types of fabrics used in the shade with no problem, as long as both fabrics are the same thickness. If you mix fabrics of different thicknesses, they will roll at different rates, and the shade will not roll up properly.

In summary, it is a good idea to have a decorative, more expensive fabric on the room side, and a less expensive, plain fabric facing out the window, as long as they are of the same thickness.

Another economical way to make an attractive shade is to buy relatively cheap quilted material, and cover it with attractive, unquilted material. We have used designer fabrics to cover quilted material in a few cases, and the results are extremely nice. You have to be very careful in picking the designer fabric to be sure it fits the window, but the end result is worth the effort.

Photo 4-1 shows a designer-print fabric on a shade that covers two small windows in an older home. To make this work, the roller box was made for a material one thickness classification thicker than the quilted material. Then the designer fabric was sewn to the face of one piece of quilted fabric, and a piece of cheaper fabric was sewn on the face of the other piece of quilted fabric. Before sewing the fabric onto the quilted fabric, the quilted fabric should be completely sewn and installed on the dowel rods. Then the designer fabric can be aligned and attached. You have to attach a piece of fabric to the other roller to keep the thicknesses of the two rollers the same, or they won't roll evenly.

Photo 4-1 — A designer-print fabric on a shade covering two small windows makes a very attractive addition to a room.

Sewing the designer fabric to the shade is difficult. Depending on the type of fabrics you are using, you'll have to stitch it at different intervals to prevent wrinkling and sagging. Begin by sewing the designer fabric to the shade on all four sides, being sure to keep it tight when sewing. To do this, first sew the top, then sew the two sides, working from the top to the bottom, then sew the bottom last. You may even have to go back and sew the middle several times. Some fabrics may have to be sewn at intervals as close as every four inches to prevent wrinkling and sagging. Attaching a designer fabric to the quilted material is a chancy operation until you learn how your fabric behaves. Ask for advice at the store where you are buying the fabric, being sure they understand how it will be used. If they can't give you much help, go to a local drapery specialist and ask for help. Designer fabrics are expensive; be sure you know it will work before buying.

One last point of concern when picking your fabric: The sun will do a real fade number on most fabrics. Even though the shade is designed to have the Mylar layer moved to the outside in hot seasons to reflect sun and heat, and the shade should be all the way up on south-facing windows on winter days, the fabrics will begin to fade in time. A light-colored, plain, quilted fabric is considered best for the outer layer of the shade. Dark fabrics will quickly fade and become uneven. A light fabric will fade somewhat, but this will be less noticeable. Spend some time in the fabric department looking at everything that is available.

The material that we have found to work the best is known as place mat material. This is made with two layers of fabric with a foam lining inside. The foam linings vary in thickness, from ⅛ inch to slightly over ¼ inch. These fabrics are easy to sew, wash nicely, and roll extremely well as a shade. Normally their quilting pattern is quite uniform, and they make very attractive shades.

We've made shades out of a wide variety of materials, ranging from 1/16-inch-thick quilted taffeta fabric to ski jacket material with almost ½ inch of lining. The thermal properties of these different materials are summed up in chart 3-1 (page 32). The thing you should consider when selecting a material, is that the thicker the material you select, the larger the roller box will need to be. However, in most cases, a thicker material may only add an extra inch to each dimension of the roller box. In many cases, the thicker materials are well worth the slightly larger roller box and higher cost.

All factors considered, we feel that ¼-inch-thick place mat material is the best all-around fabric for an insulating shade.

When picking out your fabric, inspect it thoroughly before buying. Quilted materials have a tendency to have pulls or sewing and weaving flaws. Check both sides of the fabric for flaws. If you find any, cut your fabric from a different part of the bolt.

When buying your fabric, be sure to get specific washing and drying instructions. If the fabric clerk does not know how to wash the material properly, contact the manufacturer for guidelines. Once you know the recommended washing procedures, wash the material before doing anything else to it. BE SURE YOU WASH THE MATERIAL THE SAME WAY YOU WILL BE WASHING YOUR FINISHED INSULATING SHADE IN YEARS TO COME. If you wash the material in cold water before cutting and sewing, and then wash the finished shade in hot water, it will have additional shrinkage. The material must be preshrunk before being sewn or having the fastening snaps attached, or the first time you wash the shade it will shrink and will not fit the window properly.

Many of the foam-lined quilted fabrics should not be dried in hot dryers. Air-dry them, or use a dryer at the low setting.

IN HOT DRYERS, THE FOAM MAY BEGIN TO BREAK DOWN AND MELT, RUINING THE FABRIC, YOUR SHADE, AND MAYBE THE DRYER.

When buying your fabric ask if they have plain companion fabric to go with your quilted material. Many of the place mat materials also have identical unquilted or companion materials available. These materials can be used for drapes, curtains, or even for covering the roller box, depending on how your room is decorated. Often the store will also have companion fringe and edging materials. Some may even have matching wallpaper.

The amount of fabric you will need is noted on the data sheet. REMEMBER THAT THIS IS ONLY FOR ONE PIECE OF THE SHADE. A COMPLETED SHADE NEEDS TWO PIECES OF FABRIC. Double the length measurement on your data sheet be-fore going to the fabric store.

If you decide to make both sides of the shade from the same fabric, you'll find it best to cut the two pieces to the rough length stated on the data sheet before washing. A single, long piece of fabric may get tangled in the washer, while two smaller pieces won't have that problem.

With the material washed and dried, lay it out flat on a clean floor or rug. Spend a few minutes looking at the fabric and studying how square the quilted pattern and the printed pattern are. If the two are in good alignment, no problem.

Normally, a woven pattern will be much more "in square" than a printed pattern, but the quilting process may pull any pattern out of alignment.

Study your fabric and decide what you want to have in alignment with the sides of your window, the quilting or the printed pattern. Once you decide, proceed to cut the fabric to the size on the data sheet. We suggest that you first trim one side, then cut the other long side to the width measurement. Then square off one end, and leave the other end rough. You will only be sewing three sides at this time. The fourth edge will be finished after the fabric has been installed on the rollers and the roller box is in place.

You must have some way to finish off the outside edges smoothly. We have found that the best thing to do is simply bind the edges with bias tape, as shown in photo 4-2. If you were to fold the edges over and sew a hem, the double layer of padding would make the shade roll unevenly. We have tried removing

Photo 4-2 — Binding with bias tape is an easy way to finish off the edges of the fabric.

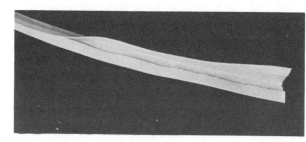

Photo 4-3 — Double-fold bias tape.

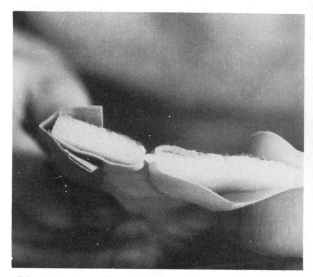

Photo 4-4 — Double-fold bias tape being sewn on the edge of a piece of fabric.

the foam and sewing a hem, and that works well, but takes a great deal of time and patience. Bias tape works very well and is quick and inexpensive.

Be sure you understand that you will not be sewing the two pieces of fabric together. Each piece remains separate. All you have to do is cut the fabric to size and sew the edging material on each piece.

There are a number of bias tape products available that will work well for the shade. We have found that ½-inch foldover tape works best. This is sometimes referred to as double-fold bias tape. It should look like the material in photo 4-3.

With the double-fold, or foldover, tape, all you need do is press the fold of the tape firmly against the edge of the fabric, as shown in photo 4-4, and sew. Be sure you are

sewing through both layers of the tape, and that you don't bunch up the tape or the fabric as you sew. Keep the tape stretched tight and against the edge of the material.

Get a color tape that closely matches your fabric, although that is not too important. The edges of the shade never show; they are always behind the side closers. The top never shows; it is in the roller box.

For the actual sewing instructions for the tape that you use, follow the manufacturer's recommendations on the package. All that matters is that you have the material covering the exposed edges of the fabric.

If you are using an extremely thick fabric for your shade, you may have to use a tape that is wider than ½ inch. Take a piece of sample material to the fabric store and buy a tape that will give you no less than ¼ inch of coverage on both sides of the fabric.

Quilted fabrics normally come in widths ranging from 42 to 45 inches. Many shade installations will require a piece of fabric wider than that. If so, you will need to put two or more pieces of fabric together. The important thing when piecing fabric for a shade is that the seam not be any thicker than the other parts of the fabric. If you were

just to overlap the fabrics and sew them, as shown in photo 4-5, that part of the shade would be thicker than the rest of the shade. When you rolled the shade it would roll faster at the seam, as it would have a larger diameter. This would cause the shade to roll unevenly, and it would quickly bunch up and not roll all the way up.

If you are attaching pieces of fabric, you have to remove some padding material from each piece before sewing them together. Photos 4-6 and 4-7 show the process we have found best to end up with a flat seam when attaching two pieces of quilted fabric. From each piece of fabric, you must remove about 1 inch of padding and fabric from the edge to be joined to the other piece, as shown in illustration 4-2. Note that on one piece the good side of the fabric is removed, and on the

Photo 4-5 — This type of seam will bunch up and roll unevenly, due to the double thickness of padding.

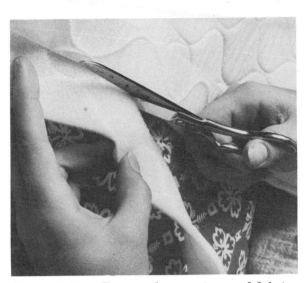

Photo 4-6 — To attach two pieces of fabric, first remove a 1-inch-wide strip of padding and one layer of fabric from each piece of quilted fabric.

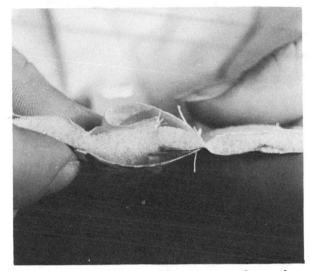

Photo 4-7 — With padding removed, overlap the two pieces of fabric and sew them together.

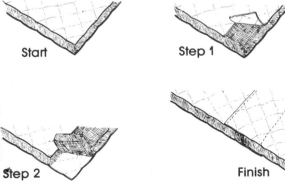

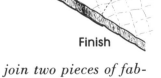

Illustration 4-2 — To join two pieces of fabric, first remove one layer of fabric and lining material from each piece. Then join the two pieces, and sew them together.

other piece, the back of the fabric is removed. Then one piece of fabric is positioned over the other piece and the two are sewn together. This leaves a strong seam that is almost as flat as the rest of the fabric.

When two or more pieces of fabric are required, be sure to put them together first, before cutting any pieces to the finished width. You won't be able to measure the width accurately until the pieces are joined.

One last point to consider when attaching two pieces of fabric: A vertical seam is stronger than piecing the two with a horizontal seam. If you make a horizontal seam, in time the stitching will begin to sag, especially toward the middle of the shade where there is little support and where all the weight of the bottom half of the shade would hang on the stitching of the seam. With a vertical seam, the two pieces tend to hang independently, and the stitching only serves to hold the two together.

At this point, you should be ready to buy your fabric and do the little sewing that it requires. As we noted earlier, proper fabric selection is harder than the actual sewing of the shade. With the sewing done, it's time to cut out and assemble your roller box, as explained in the coming chapters.

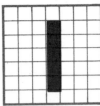 n this chapter we will explain how to cut every piece in the roller box for both a wall-mounted and a window-mounted shade. Up to this point you have measured the window and calculated the measurements for all the pieces in the roller box. You also should have bought the fabric and sewn it. Now you are ready to cut out the roller box for your shade.

Work closely with both the data sheet you have filled out for your window and the exploded illustration of the roller box on blueprint sheet 1 or 2. The combination of the data sheet, the exploded drawing, and these instructions should enable you to make each piece with no mistakes.

This chapter is in three parts. First we have the general information about building a roller box. This refers to both wall-mounted and window-mounted roller boxes. Then we present instructions for cutting a wall-mounted roller box. Last, we have instructions for cutting a window-mounted roller box. You should read the first part of the chapter and then read only the part that applies to the type of shade you are building. If you read the section that does not apply to your type of shade, you will get confused.

At the end of this chapter you should have all the pieces of your roller box cut out and ready for assembling, the subject of the next chapter.

Before you can start cutting out the pieces of your roller box, you have to buy lumber. We recommend that you make your roller box out of ¾-inch-thick material. This should be the actual measurement of the wood. To get that measurement in common lumber, ask for a 1 × 6, or a 1 × 10. The finished thickness of 1-inch nominal size lumber is ¾ inch. If your material is other than ¾ inch, you should have used the passages in italics in chapter 3, "Sizing the Shade." They explained how to size the wood for any thickness material other than ¾ inch.

However, wherever possible, we recommend you use plywood for your roller box. Common lumber will warp and bow, as shown in photo 5-1, while plywood will not have this problem. If you want to stain your roller box to leave the natural grain of the wood exposed, we recommend you buy veneer plywood. A full 4 × 8-foot sheet of veneer plywood will make three fairly large roller boxes and will be cheaper than buying that much dimensional lumber.

There are very few places where the end grain of the plywood will show. These spots can be taped with a veneer tape and stained so they will not show. Plywood is an easy and inexpensive way to make a very attractive roller box. We cannot emphasize enough how highly we recommend that you use plywood for the roller box. It is cheaper and will make a better roller box. If you want a middle ground, use plywood for all pieces that will not show and regular # 1 clear or # 2 lumber on pieces that are exposed.

There are three types of plywood available. The most expensive is known as solid core. This uses pieces of solid lumber on the inside, with a veneer plywood on the outside.

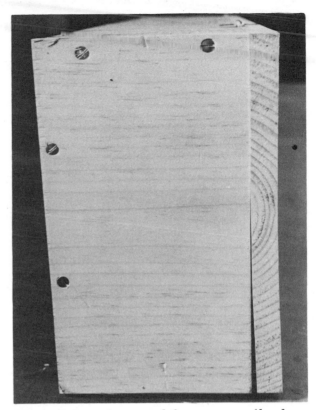

Photo 5-1 — A warped front on a roller box makes the box hard to close. Standard lumber warps easily.

One side only is veneered with a good-quality wood. Plywood with two good veneer faces is available, but very expensive. Solid core plywood gives a very good nailing or screwing surface and there is no need to use veneer tape on the solid wood edge. Ply core is standard plywood. This is strong and works well, but the edges have to be taped, and the roller box should be glued and nailed together for added strength. The last type of plywood, composition core, is the least desirable. It has poor nailing and screwing qualities. If you have to use this material, be especially careful, and do a good job of gluing the box together.

Any of these plywoods are available in veneers. Specify A-D plywood, giving you a very good surface on one side and a poor surface on the other side. If the box is going to be painted, you can go down a grade and use B-D plywood.

The most economical way to build a roller box is to combine both regular plywood and veneer plywood. On a wall-mounted shade, only the sides and the front need to be cut from veneer plywood, while on a window-mounted roller box, only the front needs to be cut from veneer plywood. By selectively mixing the types of plywood, you are using the most expensive type only where you will be benefiting from its superior qualities.

There is no way we can give you a materials list of how much wood you will need. That depends on the size of your roller box, which depends on the size of your window and the thickness of the fabric you buy.

Illustration 5-1 shows the cutting layout on a sheet of plywood for three wall-mounted roller boxes for windows 3 feet wide and 5 feet high, using a fairly thick fabric. Note how the pieces are laid out so that the grain runs the same way on all horizontal pieces and the same way on all vertical pieces. We've numbered the pieces to help you iden-

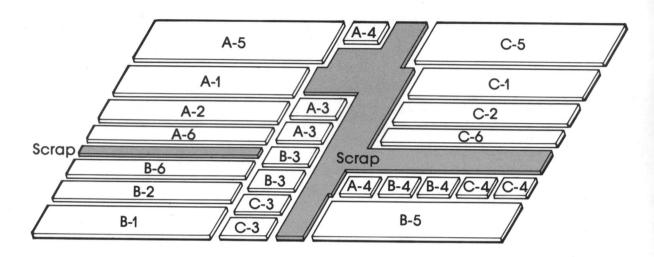

Illustration 5-1 — Sample cutting layout on a 4 x 8-foot sheet of plywood for three wall-mounted roller boxes.

tify how they are laid out. With this method, three roller boxes can be made from one 4 × 8-foot sheet of plywood. Depending on what size windows you have, you may or may not be able to use this as a guide. The idea is to plan ahead and you'll be able to save a lot on materials.

For now, look at the exploded drawing of the roller box on blueprint sheet 1 or 2, depending on which type of roller box you are making, to get an understanding of all the pieces in relationship to each other. All the pieces are cut individually. There are no joints or bevels to make.

Before cutting any wood, find the piece of wood, or section of veneer plywood, that has the most attractive grain. Save this piece for the front piece of your shade. On both

types of shades, the front is the most obvious piece, and if you are going to stain the wood, it should have the most pleasing grain pattern. Plan ahead, and your finished shade will look much nicer.

The roller box is made in two sections. First we will discuss the cutting of the pieces that go into the back unit, then the pieces in the front unit. The two units are assembled separately and then hinged together to finish the box.

For now, let's begin making a wall-mounted roller box. Get blueprint sheet 1 and your completed data sheet, and start cutting the individual pieces. If you are making a window-mounted shade, skip this section and go to the instructions that follow for a window-mounted shade.

WALL-MOUNTED SHADE

PIECE # 1 — BACK

The back is completely hidden and should be made from straight, warp-free wood. Plywood is ideal, as it is never seen and will strengthen the entire roller box. You can use a plywood back and clear lumber for other parts, and the shade will look just as good as if all the wood were of top quality. Even if you are using veneer plywood, a cheaper grade of plywood can be used for the back without hurting the appearance of the shade, saving you still more money. In the illustration, the back piece is shaded. It will be fastened to the top, the sides, and the rod holders.

The back serves to hold the entire shade to the wall and to support the sides and top. It is probably the most important piece to have right if you want a good, square roller box. The piece is simply cut as a rectangle, using all straight cuts. Be sure your blade angle is square and that the corners are square. Have the grain running the length of the piece if you are using regular lumber.

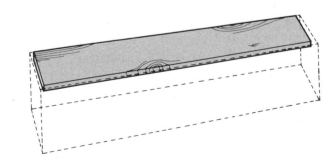

PIECE # 2 — TOP

Although it is an important piece, the top is not visible when the shade is installed and should be made from plywood to cut costs and to strengthen the unit. The top has several functions; it serves to give overall lateral strength to the roller box, and it provides an edge on which to hinge the front unit, but, most important, it prevents the formation of convection currents along the window. Without a top, warm air in the space between the shade and the window would rise or fall, depending on the season, pulling room air into the space and setting up a convection current that would cool the room in winter and heat the room in summer. But with a top on the roller box any possible convection currents are prevented from forming. The illustration shows the top in relationship to the other pieces of the roller box.

The top is cut as a straight rectangle, according to the dimensions on the data sheet. Be sure the angle of your saw is straight and that the corners are square. As with the back, if the top is not square, the entire roller box will be off.

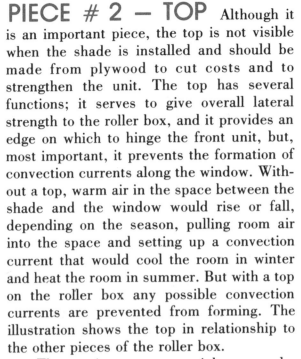

PIECE # 3 — SIDES

There are two sides for a wall-mounted shade. Each is highly visible and should be cut properly from wood with a pleasing grain pattern, especially if the unit is to be stained. If you make the sides out of plywood, you will want to use veneer tape on the exposed end grain to improve the appearance of a stained unit. If you are painting the unit, fill and sand the holes.

The sides are made simply with four straight cuts. Be sure you have the measurement running the right way with the grain. The illustration shows the sides in relation to the other pieces to which they will be attached, and their grain direction. Be sure your cuts are straight and square and the sides will cause little trouble.

PIECE # 4 — ROD HOLDERS

You will need to refer to the drawings of this particular cut on blueprint sheet 3. It's a tricky one. The measurements on the data sheet for the rod holders are for the pieces of wood from which the rod holders are made. There is a total of four pieces, that is, two pairs of pieces, in each roller box. The first step is to cut two pieces of wood to the measurements on your data sheet. If you are using regular wood, be sure you have the grain running in the direction indicated on blueprint sheet 3 and the illustration. The grain direction does not matter if you are using plywood.

The second step is to mark the center of the width of the board at both ends, and draw a line down the center of the board. Measure one-half the center-to-center distance from each end, and make marks. These will be the center points for a drill. Drill a hole for the dowel rods at each of the two center points. The hole should be the same size as, or $1/16$ inch larger than, your dowel size.

Following the drawings on blueprint sheet 3, draw lines from the edge of the drilled holes to each edge of the wood, as shown. Then connect the two holes with an angled line. Using a saber saw or coping saw, cut along the lines, dividing the piece of wood in two. Mark the two pieces piece A and piece B, making sure you have the correct label on the right piece.

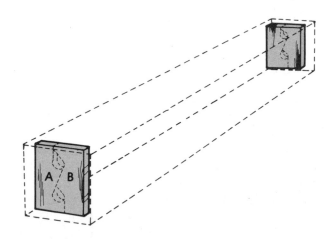

Last, trim off the two small edge pieces, as shown in the drawing, using a 45-degree angle. The two finished pieces will be used as a pair in the roller box, and should be kept together. Mark one set of rod holders L for left side and one set R for right side. It doesn't matter which side they go on as long as the pair stays together.

Repeat this process for the other piece of wood. When finished, you should end up with four pieces of wood, making two pairs of rod holders.

PIECE # 5 — FRONT

This is the most visible piece in the shade. It should be made from wood with a distinctive grain if you are staining the unit. If you are painting the unit, make it from a piece with few or no knots to give a smooth appearance. If you make the front out of veneer plywood, the roller box will work better, and the exposed edges can be taped before staining or painting. This is also the biggest piece in the roller box, so be especially careful not to make any mistakes.

The front is an easy piece to cut; it is just a rectangle with four square corners. Be sure your cuts are straight and you should not have any problems.

The illustration shows the front (shaded) in perspective to the other pieces to which it

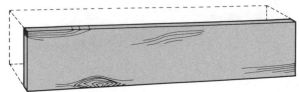

is fastened. This is the main piece of the front unit; it is also where the front unit is hinged to the back unit after assembly.

PIECE # 6 — BOTTOM

The bottom, another hidden piece, can be made of plywood and will never show. This is a simple piece to cut, a basic rectangle as shown in the illustration. Cut it to the measurements on the data sheet, and you'll be set.

Once the piece is cut, you have to drill two holes for the draw cord to pass through. These holes should be directly below the center of the pulleys when the roller box is closed. This cannot be done until both units of the roller box are assembled and hinged together.

For now, all you have to do is round the edge on one long side of the bottom board. This can be done with a rough file, coarse sandpaper, or any other device to shape wood. Don't overdo it when rounding this edge, just take off the sharp corners. Once it is shaped, go over the rounded edge with fine sandpaper to remove any splinters.

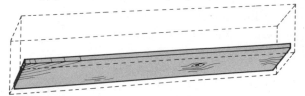

PIECE # 7 — DOWEL RODS

The dowel rods are two of the integral pieces of the shade.

Select dowels that are as straight as possible. The roller box calls for 1-inch-diameter dowels. Some lumberyards carry these in lengths up to 4 feet, but most carry only 3-foot lengths. If you need a dowel longer than 3 feet, you'll either have to buy round stock or try tool handles; as a last resort, go to a special drapery outlet. We have found that tool handles work very well for a shade. In most cases they are cheaper and straighter than dowels. Round stock comes in both 1-inch and 1¼-inch sizes; either will work for the shade. Check: If you have narrow windows you should be able to get both dowels out of one tool handle. The measurements we've given for sizing the shade will work with 1⅛-inch or even 1¼-inch dowels.

At this point, all you have to do is cut your dowels to the length indicated on the data sheet. During assembly you will be fastening pulleys to the dowels, and later you will be fastening snaps to them, but for now, just cut them to length.

PIECE # 8 — PULLEYS

Each roller box needs two pulleys, and each pulley is made of three pieces of plywood. The four side pieces are all cut to one size and the two centers to another size.

The four side pieces should be made from ⅛-inch plywood. Scrap pieces of paneling will work fine. The center pieces should be made from ⅜-inch plywood. If you use thicknesses of wood other than what we recommend, the assembled pulleys and dowel rods will not fit the fabric and the roller

box as already calculated, so follow the size recommendations exactly: ⅛-inch plywood for the four side pieces and ⅜-inch plywood for the two center pieces.

The data sheet gives the diameter of the circles comprising the pulleys. Each pulley circle needs a hole drilled exactly in the center. The hole must be exactly the same size as the dowel you are using for your shade.

It is easiest first to mark the centers of the four side pieces, then use a compass to draw the diameter of the outer circle. Then do the same for the center pieces, remembering to change the compass setting. With the center point and the outer diameter marked, you are ready to cut.

We have found it easiest to drill the center hole in all the pieces first. A hole saw is best, but spade bits will work.

With the holes drilled in each piece, the next step is to cut out the circles forming the pieces. Cut the circle carefully, especially the centers, and your finished shade will operate smoothly. If you cut the circles unevenly, the shade will roll awkwardly. After the pieces are cut, use sandpaper to smooth the sides of all cuts, especially the four side pieces. Any rough areas on these pieces may make for excess wear and tear on the pull cord when the shade is assembled. When the pieces are smoothed, put them aside; they will be assembled during the assembly stage.

That completes the cutting of the pieces for your wall-mounted roller box. Proceed to chapter 6, "Assembling the Roller Box," to see how to assemble the pieces you just cut.

WINDOW-MOUNTED SHADE

This part of the chapter details the cutting of all the pieces for a window-mounted shade. You should be using a window-mounted shade data sheet, or something is wrong. If you want to make a wall-mounted unit, use the previous part of this chapter. The two roller boxes are different, and one sized for one method of mounting cannot be mounted the other way. In fact, the measurement for one will not make the other style, as they are assembled differently. Be sure you want a window-mounted roller box, and then proceed to follow the instructions.

PIECE # 1 — BACK

The back is best made from plywood, as that will not warp and will add strength to the overall roller box without detracting from its appearance. The back is a hidden piece, so use the cheapest type of plywood you have.

In the illustration, the back is shown shaded. It is attached to the top and the rod holders. Cut this piece with extreme care. Any errors in cutting the back will show up

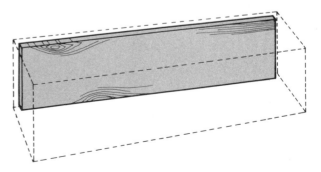

later, when you are trying to put together a square roller box. Be sure your corners are square and your cuts are straight. Have the grain of the back running lengthwise if you are using clear lumber.

PIECE # 2 — TOP
The top is an important piece, so cut it carefully. It is not seen, so the quality of the wood is not important as long as it is warp-free. This piece is a simple rectangle, cut to the measurements on the data sheet for the window you are working on. The illustration shows the top in relationship to the rest of the box.

Be sure that the cuts are straight and the angles square, and there is little that can go wrong when cutting this piece. One edge of the top will be exposed on a window-mounted shade. If you make the top from plywood, you'll have to tape this edge with veneer tape before finishing the roller box. If you will be painting your box, all you have to do is fill the holes in the plywood edge with putty, and sand and paint.

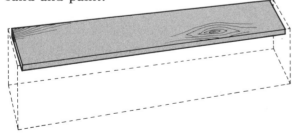

PIECE # 3 — ROD HOLDERS
The measurements on the data sheet for the rod holders are for the piece of wood from which each pair of rod holders is fashioned. Cut two pieces of wood to the measurements on the data sheet, making sure you have the grain running in the direction indicated in the drawings on blueprint sheet 4.

With one piece of wood in front of you, measure in from the left side one-half the center-to-center distance, plus ⅜ inch. Mark this at the top and bottom of the board, and draw a line the length of the board. Measure down from the top one-half the center-to-center distance, and make a center mark on the line. From that mark, measure down the line the center-to-center distance, and make a second mark. This second center mark should be one-half the center-to-center distance, plus 1½ inches, up from the bottom.

Next, drill the two holes on the center marks. The holes should be the same diameter as, or ¹⁄₁₆ inch larger than, the dowels you are using. When drilling, be careful not to drill the hole all the way through from one side. Start on one side, and as soon as the drill breaks the surface on the other side, turn the wood over and finish the hole from that side. This will prevent splintering the wood, especially with plywood.

After drilling the holes, draw a line from the edge of the board to the edge of the hole, as indicated on the drawings on blueprint sheet 4. Then draw a line connecting the two holes, as shown in the drawings. Cut the board into two pieces along the lines just drawn, and mark them piece A and piece B. After separating the two pieces, use a 45-degree angle to mark the small cut on the bottom half of piece B. At the top of the

piece, measure ⅜ inch from the cut edge, and draw a line to the outside edge of the hole. Carefully cut off these two small pieces. If these two pieces are not removed, the roller box will not close and will not work.

Mark each pair of rod holders so they will stay as a pair. They can go on either side of the roller box, but they must be used as a pair. Put the marked pair aside, and repeat the procedure on the other piece of wood.

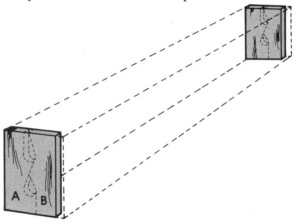

PIECE # 4 — FRONT
On a window-mounted shade, this is the only piece that shows, so make it look as nice as you can. This piece fills the entire window opening, so pick wood that is free of knots and has a pleasing grain pattern.

The piece itself is shown in the illustra-

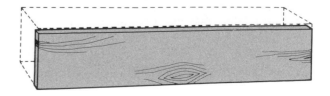

tion in contact with the other piece to which it is fastened. If it is made good and square, the finished roller box will be square and will open and close smoothly. Note the grain direction in the illustration. Keep the grain running in this direction, whether you are using lumber or veneer plywood.

PIECE # 5 — BOTTOM
The bottom piece of the window-mounted roller box is just a rectangle cut to the measurements on your data sheet, keeping the grain running the proper way. Round one of the long edges with some type of shaping tool. The edge does not have to be perfectly round, just approximately round. Sand the rounded edge smooth, and you are done with the piece for now. Later you will be mounting hardware and drilling the draw cord passage holes, but for now, you are done.

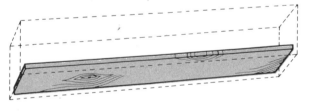

PIECE # 6 — DOWEL RODS
The dowel rods are two of the integral pieces of the shade. Select dowels that are as straight as possible. The roller box calls for 1-inch-diameter dowels. Some lumberyards carry these in lengths up to 4 feet, but most only carry 3-foot lengths. If you need a dowel longer than 3 feet and can't find it, look for either round stock or straight tool handles. As a last resort, go to a custom drapery manufacturer to buy longer dowels. We have found that tool handles work very well. They normally come in 1-inch and 1⅛-inch diameters. In most cases they are cheaper and sometimes straighter than dowels. Round stock comes in both 1-inch and 1¼-inch sizes, normally up to 6 feet long.

If you have fairly narrow windows, check; you may be able to get both dowel rods from one length of tool handle or round stock, saving some money. The measurements given for sizing the shade will work with 1⅛-inch or even 1¼-inch dowels.

At this point, all you have to do is cut your dowel to the length indicated on your data sheet. You'll be doing a lot more with the dowel rods during the assembly stage.

PIECE # 8 — PULLEYS
Each roller box needs two pulleys, and each pulley is made of three pieces of plywood. Thus, each roller box needs six pieces of plywood to make the pulleys.

The four side pieces must be made from ⅛-inch plywood. Scrap pieces of paneling work just fine. The two center pieces must be made from ⅜-inch plywood. If you use thicknesses of wood other than what we recommend, the assembled pulleys and dowel rods will not fit with the fabric in the roller box, so follow the size recommendations exactly: ⅛-inch plywood for the four side pieces and ⅜-inch plywood for the two center pieces.

The data sheet gives the diameter of the two sizes of circles needed for each pulley. Each circle needs a hole exactly in the center. The hole should be exactly the same size as the dowel you are using for your shade.

It is easiest to mark the centers of the four side pieces first, then use a compass to draw the diameter of the outer circles. Then do the same for the center pieces, remembering to change the compass setting. With the center points marked and the diameters marked, you are ready to cut and drill the wood.

We have found it easiest to drill the center hole in all the pieces first. The hole size is the same for all the pieces, so you only have to set up your tools once. A hole saw works best, but standard spade bits will also work. Be careful to avoid splintering the plywood when the drill breaks through.

With the holes drilled in each piece, the next step is to cut out the circles forming the pieces. Cut the circle carefully, especially for the two center pieces. If these circles are too far out of line, your finished shade will not roll up and down smoothly. After the circles are cut, use sandpaper to smooth the sides of all the pieces. Pay special attention to the side pieces, for any splinters or rough spots on these will rub against the draw cord, creating excess wear and tear on the cord during use. When the pieces are smoothed, put them aside; they will be assembled later.

That completes the cutting of the pieces for a window-mounted roller box. The next chapter details how to put the pieces together into a finished roller box.

6 ASSEMBLING THE ROLLER BOX

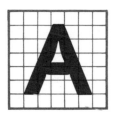t this point all the pieces for your roller box should be cut and awaiting assembly. Now you'll find out if your cuts were straight and square and if your measurements were accurate.

This chapter is broken into three parts. First there is general information about assembling a roller box. Then there are specific instructions for assembling a wall-mounted roller box and last, specific instructions for assembling a window-mounted roller box. Everyone should read the first section, and then only the section that applies to the type of roller box being built. The two types of roller boxes are put together differently, so be sure you read the appropriate part.

The roller box is assembled in two sections, a front unit and a back unit, and those sections are hinged together, forming the completed box. The back unit is made up of the back, the top, the sides (wall-mounted roller box only), and piece A of the rod holders. The front unit is made up of the front, the bottom, and piece B of the rod holders. The dowel rods and pulleys comprise a separate unit.

The method of putting the pieces together is quite easy; all joints are simple butt joints, glued and nailed. One joint is also screwed together on a window-mounted unit to give additional strength. We recommend you assemble the roller box with finishing nails. On all pieces that are exposed, the finishing nails can be countersunk and the holes filled with putty.

If you use plywood to build your box and plan to stain the finished unit, you should put veneer tape on all exposed edges before putting the unit together. If you put the tape on after the unit is assembled, the tape will show. By taping first, the chances of getting a smooth joint are much greater. Illustration 6-1 shows the edges that need to be taped on either a wall-mounted or a window-mounted plywood roller box.

There are two types of veneer tape available; one has heat-activated adhesive on it, the other requires the use of separate adhesive. Both types work well and are easy to use. Follow the directions for whatever type of tape you get, take your time, and the results should be quite good. If properly applied, stained, and finished, the tape will last a very long time.

With all exposed edges taped, you are ready to put the pieces together. Nailing butt joints together can be somewhat tricky—the pieces always seem to want to jump out of alignment just as the nail starts to penetrate. We have found that it helps to drill pilot holes through one piece of wood before nailing.

When nailing into plywood you will get a tighter joint if the nail goes in at a slight angle. Illustration 6-2 shows a butt joint; note the pilot hole alignment line drawn ¼ inch from the edge, and a modest angle for the pilot hole. If this angle is too great, the nails will split through the inside face of the second piece. Note: Only drill the pilot hole through the top piece of wood. Select a pilot hole size that will still give some bite to your nails when putting them through the wood. We have

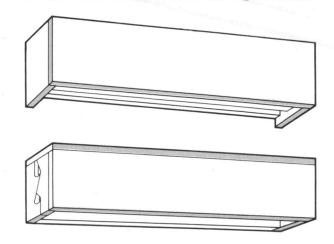

Illustration 6-1 — Plywood wall-mounted and window-mounted roller box edges that need veneer tape are shown shaded.

found 6d or 8d finishing nails are best for putting the box together and a ¹/₁₆-inch pilot hole is best for that nail.

Standard woodworking glue is fine for gluing the pieces of the roller box together. Always spread a thin coat of glue over the entire length of the joint after drilling the pilot holes. Let the glue set for a minute or two, and then nail the pieces together. Be sure the glue is still wet to the touch when putting the pieces together. If you are careful with the glue when putting the box together, you shouldn't end up with glue stains all over the box. Stains from excess glue have to be sanded later, or the glue can be wiped off with a wet cloth before it dries.

Next we will explain how to put the dowel rod/pulley assembly together. Everyone should read this, as the procedure is the same

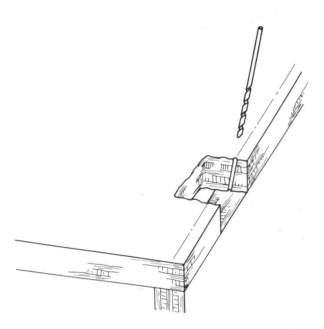

Illustration 6-2 — Butt joints are easier with an angled pilot hole through the top piece of wood.

for both types of roller boxes. Then we will explain a wall-mounted roller box assembly and last, a window-mounted roller box assembly. Be sure you follow the right set of instructions, as the boxes are put together differently.

DOWEL ROD/ PULLEY ASSEMBLY

It does not matter which type of shade you are building; the dowel rod/pulley assembly is the same for both shades. This is a somewhat tricky stage of assembly, so follow the directions exactly.

It is very important that the three pulley pieces be assembled on the dowel rod in exactly the correct spot. The pulley pieces must also be positioned as close to perfectly

perpendicular to the dowel as possible.

Begin the assembly stages by cutting a 3- or 4-inch piece of scrap dowel. This piece will be used to align the pulley pieces for assembly. Be sure the dowel has good square ends.

Position the scrap piece of dowel on end, on a flat surface. Slip a pulley side, center, and side on the dowel, to check the fit. Then remove them, and apply a light layer of glue to one side of each side piece and both sides of the center, as shown in illustration 6-3.

Restack the three glued pieces on the dowel, checking alignment. When the pieces are lined up, nail the pieces together with ½-inch brads. Nail from both sides to hold the pulley firmly together. Once the pieces are tacked together with the brads, remove the unit from the dowel and set it aside to dry. Repeat the procedure for the other pulley.

To fasten the two assembled pulleys to the dowel rod, first measure $13/16$ inch from one end of the dowel rod in several places and make pencil marks. This will be the line for the outside edge of the pulley. It allows ¾ inch for the dowel rod to fit into the rod holder and $1/16$ inch for clearance. Mark the alignment line on several sides of the dowel rod to help in the gluing stage.

If you are making your roller box from material other than ¾-inch stock, measure in from the end of the dowel rod the thickness of the rod holder, plus $1/16$ inch.

Check the fit of the pulley assemblies on the dowel rod. If they fit snugly, use standard woodworking glue to hold them in place. If they fit loosely, we recommend you use an epoxy-type of glue for better bonding. Apply

Glue surfaces

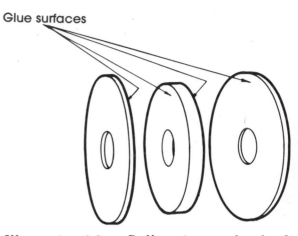

Illustration 6-3 — Pulley pieces to be glued.

a liberal coat of glue to the dowel rod, and slip the pulley into place, using the lines you've marked on the dowel rod as your guide, as shown in photo 6-1. With the pulley in place, revolve the dowel rod to see if the pulley is perpendicular; if it is, set it aside to dry. For drying without knocking the pulley out of alignment, set the dowel rod on a table for several hours with the pulley over the edge. Repeat the process for the other dowel rod and pulley.

From here on, the instructions are split. Be sure you follow the instructions for the type of unit you are building.

ASSEMBLING A WALL-MOUNTED ROLLER BOX

As pointed out earlier, the roller box is built in two sections, the front unit and the back unit.

Photo 6-1 — Final pulley position on the dowel is marked with pencil marks.

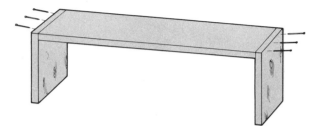

Illustration 6-4 — Attaching sides to the top.

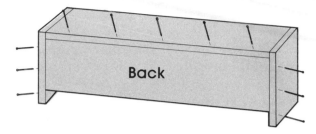

Illustration 6-5 — Attaching back to side/top unit.

First we will build the back unit. This section is made up of the back (piece # 1), the top (piece # 2), the sides (piece # 3), and the rear portion of the rod holders (piece # 4A). Pull these pieces out and set them aside.

The first step is to attach the sides to the top. Be sure you have the sides in the right position. They are rectangular pieces, and can only go in one direction. If you followed the grain pattern shown earlier, follow the grain pattern in illustration 6-4 now. Otherwise, go back and check your measurements. The sides should be exactly the same width as the top. Illustration 6-4 shows how the sides overlap the top. Do not have the top overlap the sides, or the box will be too small. Select the better side of each side piece to face outward. Once you have checked the fit of the pieces, drill pilot holes in the sides, and apply glue to the two end edges of the top

and along the top of the inside face of the sides. Glue and nail the other side piece in place the same way.

Next, attach the back to the top/side unit you've already assembled. Handle the top/side unit carefully, as the side pieces are easily knocked out of alignment at this time. The back is fastened to both the sides and the top piece. Illustration 6-5 shows the roller box from the back; you can see where the back is fastened to the other pieces. Note that the back is not the same height as the sides, and does not extend all the way to the bottom of the sides. For this reason the pieces must be put together upside down. Position the top/side unit on a work surface on the top, with the sides facing up; drill pilot holes in the side pieces; and glue and nail the back in place, so it is firmly against the top.

The last pieces to be attached to the back unit are the rear sections of the rod holders. Be sure you have the right pieces and that you are placing them correctly before gluing and nailing them. When you are sure you have the correct pieces properly positioned, as shown in photo 6-2, spread a layer of glue on one face of the rod holder and on the area of the side where it will go. Use a

Photo 6-2 — Position rear rod holder carefully.

smaller nail—a 3d finishing nail works well, larger nails will show through the face of the side. Make sure it fits tightly against the edges of the already built top/back/side unit. Again, be sure you have the piece properly positioned before gluing and nailing.

This completes the assembly of the back unit. Check it over; if any glue has squeezed out of the joints, use warm water to wash it off. Set the completed unit aside for now.

The front unit is made of the front (piece # 5), the bottom (piece # 6), and the front half of the rod holders (piece # 4B). It isn't hard to find out where these pieces go, but because of the shape of the pieces, getting them firmly in position for nailing is tricky. Using pilot holes is especially important and helpful.

The bottom and both front pieces of the rod holders must have a full ¾-inch setback when they are fastened to the front.

If you are using wood of a different thickness, you'll have to allow for it. The setback is the same as the thickness of wood used for the sides.

Decide which side of the front piece you want showing, lightly mark it as the face, and lightly draw a line for the pilot holes, 1 inch up from the edges on the bottom and both sides. Drill pilot holes along this line at a slight angle, as already explained.

Next, attach the front half of the rod holder to the bottom. There is only room for two nails, so be sure to use pilot holes for a good alignment. Note in illustration 6-6 that the bottom piece overlaps the rod holders, so the pilot holes go in the bottom piece. Glue

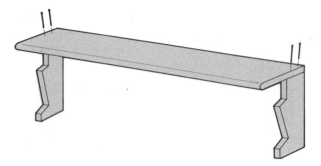

Illustration 6-6 — Front part of rod holder properly positioned on bottom piece.

and nail these three pieces together. At this stage, handle the group of pieces very gently, as the rod holders are very vulnerable to being knocked out of position.

To attach the front, first turn it over on its face. On the back side of the wood mark a line ¾ inch from the edge along the entire length of the bottom edge. This will be the alignment line for the bottom unit. Standing the front piece on end, put a piece of scrap ¾-inch wood against the side. This will act as a spacer to keep the bottom/rod holder unit set back the necessary ¾ inch from the edge. Stand the bottom/rod holder unit in place on the piece of scrap wood, and position it according to the alignment line you have drawn on the back of the front piece. When the bottom is in line with the line, and the rod holder is on the ¾-inch spacer, the pieces should be in proper position.

Take them apart, apply the glue, and let it set just a few seconds. Allow for additional time to reposition the pieces before the glue sets too dry. Reposition the pieces and check the alignment. Nail the rod holder and one end of the bottom in place, being sure the

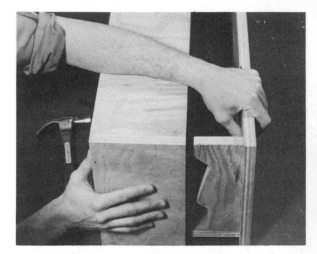

Photo 6-3 — Two units ready to mark hinge placement.

bottom stays in line with the alignment line. Flip the unit over, and nail the other end in place, again using the scrap wood as a spacer.

You may find a different way to put these pieces together that is more convenient for you; if so, go ahead. The main concern is that the bottom and the rod holders must have a full ¾-inch setback from the edges of the front piece. Be sure they are glued and nailed well. That completes the assembly of the front unit.

Now the two units must be hinged together. Stand both units on a flat surface, as shown in photo 6-6. Slide the front unit into the back unit until they are firmly together. It is most important that the top edge of the front piece be aligned with, and flush against, the edge of the top. Once the two pieces fit together well, mark where you want your

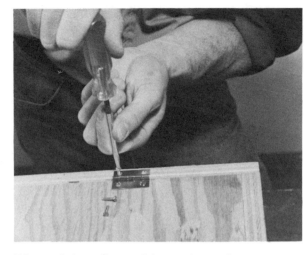

Photo 6-4 — Screw hinges into place.

Photo 6-5 — Carefully extend the center line of the rod holder holes.

Photo 6-6 — Using a square, transfer the mark to the bottom piece.

hinges. We use standard 1½-inch by 1½-inch open hinges. The hinges are simply screwed into the face of the top and the top edge of the front, as shown in photo 6-4. The hinges are hidden when the box is installed. Two hinges will be enough for most windows, although for a box over 40 inches in length, three hinges would be advisable. If the front or top are warped, extra hinges may help to straighten the wood.

With the hinges in place, test the unit to see how it swings. It may be necessary to file or sand the outer edges of the front pieces of the rod holders to get the two units to close properly. Also check to see if there is any binding between the sides and the front parts of the rod holders or the bottom. If the unit swings freely, you've done a good job. If it does not, take time to find out exactly where it is binding, and then file the area until it swings freely.

With the two units hinged together, all that remains is to drill two holes for the pull cord to pass through, and attach the closing latches. You will need to drill a ½-inch hole directly below the bottom pulley for the pull cord to pass through. The shade is designed so you can have the pull cord operate on either side of the roller box, so we recommend you drill a hole on both sides. If you are sure you won't change the side of the window you want the pull cord on, you only have to drill one hole.

To find the center location for the hole, first stand the box on end and open it up. Using a straightedge, draw a line on the bottom of the side piece to mark the center of the rod holder holes. If the marks are still on the rod holder from when you cut it, just extend the line you used to drill the holes. If the lines are not visible, eyeball the center of

the two holes you drilled and line them up with a straightedge, as shown in photo 6-5, and mark the bottom of the side piece.

With that done, close the roller box, and, using a square, as shown in photo 6-6, transfer the mark to the bottom piece. Measure in from the edge of the bottom 1⅛ inches along the line, and make a mark. This will be the center point for the hole, as shown in illustration 6-7.

Once you have marked the center location for the hole, proceed to drill the hole with a ½-inch bit. Be careful not to drill all the way through the wood from one side. As soon as the point of the drill breaks the surface of the other side of the wood, stop drilling. Use the break-through point as a guide to drill the remainder of the hole from the other side. With the hole drilled, you have to bevel the sides of the hole until it is well beveled, as

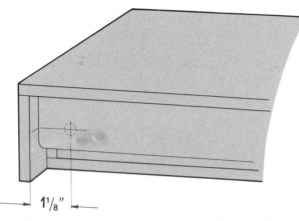

Illustration 6-7 — Marking for pull cord hole.

Photo 6-7 — Bevel the cord hole until it is smooth.

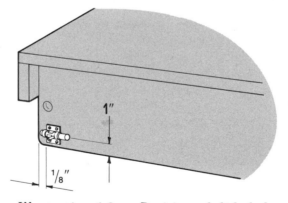

Illustration 6-8 — Position of slide bolt.

shown in photo 6-7. A large countersink will work well if you have one.

Last you need to install the closing latches. We have found that standard slide bolts work best. If you use the standard, round barrel slide bolts most often sold, you need only mount the one piece on the bottom and drill a hole in the side into which the bolt will go. However, we like to use the more decorative square slide bolts sold for cabinets. These require that you chisel a square hole in the side to receive the locking bolt. This is easily done, if you have a small, very sharp chisel and some patience. The bolts should be mounted on the bottom side of the bottom piece near the front piece, as shown in illustration 6-8. Be sure they don't get in the way of the cord hole. The slide bolt itself is mounted to the wood with the screws provided with the unit.

That completes the assembly of your wall-mounted roller box. The remainder of this chapter details assembly of a window-mounted unit. Chapter 7, "Installing the Shade," completes your insulating window shade by showing you how to install it on the window.

ASSEMBLING A WINDOW-MOUNTED ROLLER BOX

Reread the introductory part of this chapter before going any further with this section.

A window-mounted roller box is made in two sections; a back unit and a front unit. The back unit is made from the back (piece # 1), the top (piece # 2), and the rear pieces of the rod holders (piece # 3A). The front unit is made from the front pieces of the rod holder (piece # 3B), the front (piece # 4), and the bottom (piece # 5). These two units are then hinged together to form the roller box. The dowel rod/pulley assembly is a separate part, already explained.

For now, let's work on putting together the back unit. Get the back and the two rear parts of the rod holders. Illustration 6-9 shows these three pieces assembled. Note how the rod holders overlap the back. Also note the position of the rod holders. They must both be in the same position. As explained earlier, drill pilot holes for the nail to help get a good fit. Apply glue to the pieces being assembled, let the glue dry slightly, and nail the pieces together.

With these three pieces assembled, you are ready to put on the top. The top overlaps the other three pieces, so drill the pilot holes through the top piece. Put glue on all edges and areas to be joined, position the top, and nail it in place, as shown in photo 6-8.

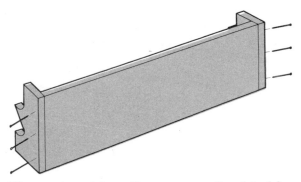

Illustration 6-9 — Rear parts of rod holders nailed to the back piece.

Once the top is firmly nailed, we recommend you go back and put several screws through the top into the back and the rod holders. There is a lot of downward pressure on a window-mounted shade at the joining point of the top to the back and sides. Screws will strengthen the joint. Three 2-inch # 10 flathead wood screws along the back piece and one into each of the rod holders will do. We recommend you drill a 3/32-inch pilot hole in the back and rod holders, and a 3/16-inch hole in the top. Countersink the heads of the screws so they are flush. Tighten the screws, but be careful not to overtighten them; with plywood it is fairly easy to strip the threads when going into the end grain. When done, set this unit aside.

To assemble the front unit, first select the best surface of the front for the face of the box. This is really the only piece of a window-mounted shade that shows, so be sure you have the best side facing the room, especially if the unit is to be stained.

Next, fasten the front parts of the rod holders to the front. Nailing these pieces

Photo 6-8 — Nail the top to the rod holder/back unit.

together is a little tricky, as you cannot get a good steady positioning of the pieces. The use of pilot holes through the front piece is especially important. With pilot holes drilled on both ends of the front, glue and nail the two rod holder pieces to the front. Be sure the front overlaps the two rod holder pieces. This is just the opposite of the way the back and the rod holders fit together, so check illustration 6-10 to be sure you have the pieces placed correctly before nailing. Both rod holders on the front must be positioned the same way, as shown. Double-check everything before nailing.

With the three pieces together, all that remains is to fasten the bottom to the just-completed front/rod holder unit. Be sure that the bottom is attached to the bottom end of the rod holders. Illustration 6-11 shows the

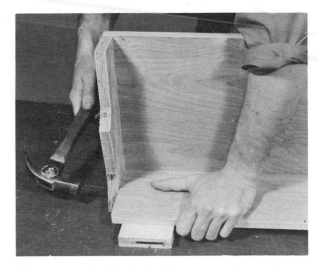

Photo 6-9 — Nail the bottom to the rod holder/front unit.

way the bottom should be placed in relationship to the rod holders. The bottom is recessed 3/4 inch from the bottom edge of the front. Draw a line on the face of the front, 1 inch from the edge, to center your pilot holes, being sure to drill the pilot holes at the proper angle. With the holes drilled, turn the unit on edge. Use scrap pieces of 3/4-inch material to position the bottom piece. Hold it tight against the inside of the front piece, as shown in photo 6-9. The bottom should fit perfectly between the two rod holder pieces. Apply glue to the edges of the bottom and the areas where it will be fastened to the front and rod holder pieces. With the bottom resting on the 3/4-inch pieces of scrap, nail the bottom to the front/rod holder unit, making a single unit of the four pieces and completing this section of the roller box.

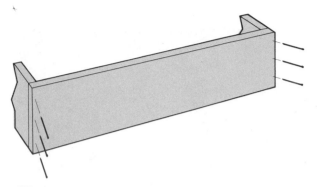

Illustration 6-10 — Front parts of rod holders nailed to the front piece.

To hinge the front and back units together, use standard 1½-inch by 1½-inch open hinges. On a window-mounted unit, the hinges must be installed between the front piece and the top piece. This requires that you cut a recess notch in each piece for the hinge to rest in. On the front, this recess is cut into the edge; on the top it is cut into the face of the bottom of the top. Be extremely careful when cutting the recesses in the top as the veneer tape can easily be damaged. Always try to cut from the tape to the wood; never cut so that the tape is the last thing you cut, as it will easily peel off.

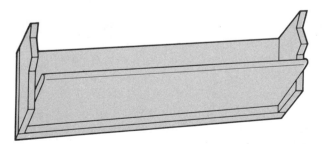

Illustration 6-11 — Bottom being fastened to rod holder/front unit.

Photo 6-10 shows the hinge recess being cut in the front unit, and photo 6-11, the recess being cut in the top. Be careful that the two recesses line up properly. You may find it easier to drill the holes for the hinge screws first, then, with the two roller box units apart, cut each hinge recess. That way, you'll know the finished units will fit. Another approach is to mark and cut the recesses in one unit, fasten the hinges to the unit, then mark the other unit and cut those recesses. If you choose this method, which we prefer, be sure to put the hinges on the top piece first, then mark and cut the front piece, as it is the easier piece to cut. The recesses should be the exact thickness of the hinge flanges that will be fastened into them. If you undercut, remove the hinge and cut some more. If you overcut, remove the hinge and shim the area with small pieces of cardboard.

When testing to see if the hinge fits properly, be careful not to overtighten the hinge screws, stripping the wood. If this happens, the screws will not hold tightly enough to work for very long, and eventually you'll have to move the hinges.

With both hinges in place, check the operation of the roller box. It should swing open freely. If not, check the edges of the rod holders; they may not be cut back quite enough and may bind. If there is a problem, carefully study the unit and find out exactly where it is hitting, then file or cut to make it operate smoothly. When finally adjusted, the unit should swing open and shut freely, with no binding or rubbing of the wood.

Photo 6-10 — Cut the hinge recess in the front unit.

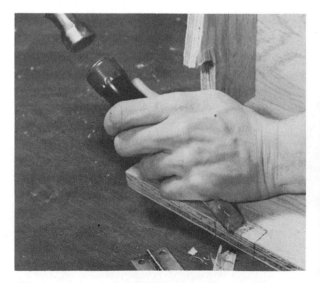

Photo 6-11 — Cut the hinge recess in the top.

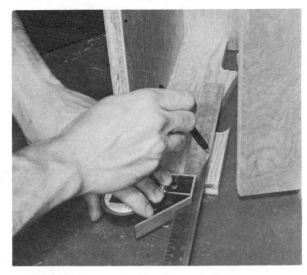

Photo 6-12 — Extend the rod holder centerline to the bottom.

Photo 6-13 — The cord hole can be extended into a horseshoe-shaped opening.

With the two units hinged together, you now have a window-mounted roller box. All that remains to be done is to cut the holes for the pull cords and attach the closing latches. First, we'll drill the pull cord holes.

Stand the box on end and open it up. Extend the centerline of the holes in the rod holder to the bottom, as shown in photo 6-12. Close the box and, using a square, extend that line across the bottom piece. Then, measure in from the side ⅝ inch and make a centering mark. This should be directly under the bottom pulley. Repeat this procedure on the other end of the box, if you want your shade to be reversible. If you are sure you won't want to change the side your pulley cord is on, mark only one side.

Using a ½-inch drill, drill from the bottom face of the bottom toward the inside of the box, on the center mark. As soon as the drill breaks the surface of the wood, stop and finish the hole from the other side. This will prevent splintering the wood. With the hole drilled, taper the edges with a countersink, a sharp knife, or a chisel.

Because this hole is so close to the edge, once it is drilled you may want to notch it out, forming a horseshoe-shaped opening, as shown in photo 6-13. This will not hurt the performance of the shade and may actually make it easier to open, as you will be able to slip the pull cord out of the groove without disconnecting it from the bottom tensioner.

The final step in assembling your window-mounted roller box is to install the closing latches. While standard, round barrel slide bolts will work, we have found that square slide bolts work and look better—but they cost more. To install round barrel slide bolts you need only to mount the bolt units, as shown in illustration 6-8 (page 64), and drill holes in the rear rod holders to receive the sliding bolts.

To install square slide bolts, install the units as shown in illustration 6-8, but this time form square holes in the rod holders to receive the bolts. This is best done with a sharp chisel. Work slowly and get a good, firm fit, not too tight and not too loose. You want the slides to operate smoothly, with about ¼ inch of the bolts penetrating the wood to hold the unit together.

With the slide bolts installed, the window-mounted roller box is completed. The next chapter details how to install the finished shade.

7 INSTALLING THE SHADE

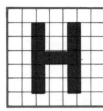ere comes the fun part, putting all the pieces together into a finished shade. With each step from here on, your work will be coming together, resulting in an attractive window shade that saves you energy and money while making your home more comfortable.

This chapter works basically the same as the other chapters in section II, covering both a wall-mounted shade and a window-mounted shade. First we will explain how to install a wall-mounted roller box and measure and build the side closers. Then we will do the same for a window-mounted shade. Last, we will discuss how to install the fabric and Mylar and make the bottom molding. The process is identical for both types of shades.

Only read the part of the beginning of the chapter that pertains to the type of shade you are making. Everyone should read the last section, which covers attaching the fabrics, adjusting the shade, and operating it.

The decision is yours whether to paint the roller box before you install it or after it is up. If you plan on staining the unit, we strongly suggest you stain it before you install it. You will get a much better paint or stain job by doing it before the box is up, but you run the risk of scraping the paint if you are not careful during installation. A compromise is to prime and paint the box before installing it, and then put a final coat of paint on the areas that show, after the box is installed.

INSTALLING A WALL-MOUNTED ROLLER BOX

The first step in installing the shade is to fasten the roller box to the wall, directly above and centered over the window. The box is designed to be centered over the window. If you took your measurements correctly and built the box to size, it should fit directly over the window. Illustration 7-1 shows the details of standard 2 × 4 wall construction for most contemporary homes. Note the use of a larger-size header on top of the window opening. This will give you a very sturdy surface on which to attach the box. You should fasten the box to a wall with two sets of screws through the back of the box. Keep one set of screws toward the bottom half of the roller box and your chances of hitting solid wood will be better. These screws should be at least 2½ inches long. A second set of screws should go at the top of the box and should go into the vertical studs.

If you find you cannot fasten into solid wood, you will have to use one of the many types of fasteners designed for hollow walls. Ask your hardware store what they think will work best. Tell them exactly what you will be hanging and that it will be used daily, so you need a very solid mounting. Any looseness or wobble will quickly get worse.

Installing the roller box is a two-person job. One person should hold the box while the other checks to be sure it is level and

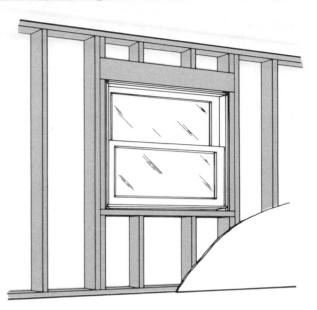

Illustration 7-1 — Standard 2 x 4 wall construction; note stud placement for fastening areas.

properly positioned, as shown in photo 7-1. Once the box is in position, one person should hold it, while the other marks and drills the holes.

Use # 10 flathead wood screws, and drill a pilot hole through the back of the box with a 3/16-inch drill bit. Only go a little distance into the wall with that drill, just enough to mark where you want the hole in the wall. Then change drill bits and drill a smaller (3/32-inch) hole into the wall. If you were to drill the pilot hole through the back of the box and all the way into the wall, it would make a hole too big for the screw to tighten in the wood of the wall. By drilling two different size holes, the larger shank portion of the screw has one size

Photo 7-1 — Two persons are needed to position the roller box.

pilot hole in the box, and the smaller screw portion has another size hole in the wall. Last, countersink the screwhead hole on the back of the box. Have your helper reposition the box, while you screw it tight to the wall. The box is not too heavy, but holding it open with the front raised is somewhat inconvenient, and it should be done by two people. The critical points in installing the roller box are that the back of the box be firmly against the wall; that the bottom of the back be sitting on the window trim; that the box be exactly centered from side to side over the window; and that the box be perpendicular to the sides of the window.

With the roller box permanently mounted on the wall, you are ready to measure and construct the side closers. Look at blueprint sheet 1 to get an idea of how the closers are made. Each side is simply two pieces of wood. The two pieces are held together with spring-loaded cabinet hinges. This enables the closers to apply a lot of pressure to the fabric, giving an almost perfect seal. Yet when the sides are opened the pressure is released, and the fabric can be freely rolled up, without any wear and tear from the side closers.

Illustration 7-2 shows how the closer looks in the closed position on a wall-mounted shade. Note how the mounting board (piece # 9) is directly against the existing window trim, and how the side closer (piece # 10) holds the fabric firmly against the trim. Each shade needs two mounting boards and two side closers.

The first step in making the mounting boards and the side closers is to measure the installation. Measure from the bottom edge of the back of the shade to the top of the sill. Be sure you are measuring from the bottom of the back, as shown in photo 7-2, not the bottom of the sides. This will give you the exact length of the mounting board.

If you made your roller box to the measurements on the data sheet, each mounting board should be 1½ inches wide. At that width, they will fit between the window trim and the roller box sides. The wood for the mounting boards should be the same thickness as the trim on the window.

With the two mounting boards cut, test-position them. If you have any abnormal trim, you will have to shave, shim, or notch the mounting boards to get a good flush fit. Once the mounting boards are shaped to fit well, proceed to make the side closers.

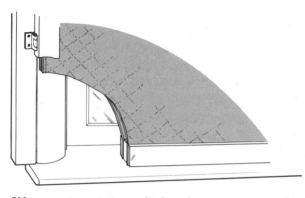

Illustration 7-2 — Side closer on a wall-mounted shade.

The closers should be 1½ inches shorter than the mounting boards. The closers can be the same thickness as the mounting boards, and should also be 1½ inches wide.

With two mounting boards and two closers cut, sand them smooth, and using either rough sandpaper, a file, or some other shaper, round the edges on all pieces. Before putting the pieces together, either stain or paint them.

The two pieces are fastened together with spring-loaded cabinet hinges. These hinges mount flush on the mounting board, and with the closer attached, form a nearly airtight seal on the trim. Assemble the two pieces, as shown in photo 7-3, being sure to keep the bottoms almost flush. The closer should have about $1/16$ inch clearance with the sill. The tops will be at different heights. There should not be more than 24 inches between hinges. Position them equally for the height of your window, as shown in illustration 7-3. Note: A hinge should be within 10 inches of each end of the closer.

Photo 7-2 — *Measure the length of the mounting boards from the bottom of the back.*

Leave about $1/16$ inch of space at the bottom of the unit, so the closer can swing freely.

Unless you are a very neat worker, stain or paint the side closer and mounting board before assembling them. Once together, or even worse, on the wall, it is difficult to finish without creating a mess.

You should attach some type of knob or handle on the closer before mounting it on the wall permanently. Don't position the knob or handle at the top or bottom of the closer, but somewhere near the middle.

With the mounting board and closer

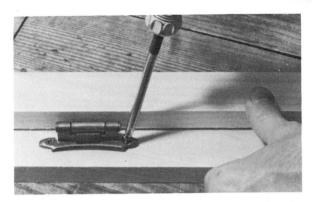

Photo 7-3 — *Join the mounting board and side closer with spring-loaded hinges.*

assembled into a single unit, all that remains is to fasten this unit to the wall. Look at illustration 7-1 (page 69) again, and see where studs are located on a typical window installation. You should be able to fasten the mounting board with nails or screws to a stud. If you cannot do this, you will need to use a wall fastener.

One of the key features of Rodale's Insulating Window Shade is that, by sealing outside of the trim, the shade reduces most infiltration through the window. To further improve on the anti-infiltration features of the shade, apply a bead of caulking along the junction of the wall and the trim. When the mounting board is put in place this way, it will be even more effective in sealing air leaks.

With the mounting boards and closers installed, you are ready to attach the fabric. The next section is for a window-mounted shade. Skip it, and go to the part on attaching the fabric.

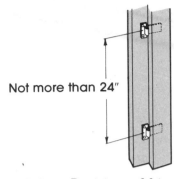

Not more than 24″

Illustration 7-3 — *Position of hinges on the mounting board and side closer.*

INSTALLING A WINDOW-MOUNTED ROLLER BOX

The first step in installing a shade is to fasten the roller box to the window cavity. Once the roller box is in place, you can measure and cut the side pieces and the fabric. Because windows vary, we recommend that each piece during this phase of assembly be individually measured and cut after the roller box is installed.

Most deep windowsills that would be using a window-mounted shade are made of some type of masonry material. Masonry creates some special fastening problems. Whenever possible, we recommend you fasten the roller box to the wooden parts of the window.

If you can't fasten to wood, you'll be forced to fasten to the other areas of the window frame. Find out what type of fastening surface you will be working with, and go to the local hardware store for advice. There are a great many fastening options. Explain exactly what your problem is, and they should be able to come up with a solution. If not, change hardware stores.

Don't forget, you can fasten the window-mounted roller box by screwing through the top of the unit, not just through the back. We have allowed only $1/16$ inch clearance between each side of the box and the walls. In some cases you will be able to screw through the rod holder and into solid wood on one side of the window jamb.

Installing the roller box is a two-person job. One person holds the box in place and keeps the front open, while the other positions it and drills the mounting holes. A window-mounted roller box is easy to position, just put it in the window opening for which it was made. There is a slight clearance on each side, and the back and top should fit flush against the window jamb.

Once the box is in position, mark where you want to drill your holes. You can either drill through the box at this time and let the drill lightly penetrate the other surface to mark where the holes should be drilled, or you can mark the box, take it down and drill holes in it, change drill bits, and reposition the box to mark where to drill into the walls.

Be careful that you know what size holes to drill and where to drill them for the type of fastener you are using. Some of the masonry fasteners require quite large holes in the masonry, and small holes in the wood of the roller box. Find out exactly what size holes

your fasteners need, and be sure to put them in the right place.

Remember, this is a two-person job. Have a helper.

With the roller box permanently mounted in the window opening, you are ready to measure and construct the two side closing units. Look at blueprint sheet 2 to get an idea of how the closers are made. Each side is made of two pieces of wood. The wood is held together with spring-loaded cabinet hinges. This enables the closers to apply a lot of pressure to the fabric, giving an almost perfect air seal. When the sides are opened, the pressure is removed and the fabric can be freely rolled up or down without any wear and tear from the side closers.

Illustration 7-4 shows how the side closing units work on a window-mounted unit. Note that the mounting board (piece # 8) is wider than the closer (piece # 9). This allows the closer to seal the fabric against the extra, exposed mounting board face, giving an almost perfect seal.

The mounting board actually can extend somewhat past the window frame, without hindering the operation of the window, allowing the shade to be installed on windows with very little mounting space. As explained in chapter 3, "Sizing the Shade," if you have at least 1 inch of space for installing the mounting board, you can install a window-mounted shade.

A window-mounted shade uses two mounting boards and two side closers. The first step is to determine the size of the mounting boards. Measure from the bottom

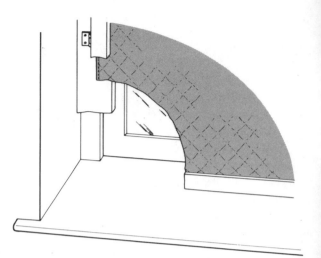

Illustration 7-4 — Side closer on a window-mounted shade.

edge of the back to the top of the sill for the length of the mounting board.

The width of the mounting boards is somewhat variable, depending on your window. For most windows, a mounting board 3 inches wide is ideal; that is the size we recommend.

The closer strips should be 1½ inches shorter than the mounting boards and ¾ inch narrower than the mounting boards. You want the outside edges of the mounting board and the closer to be flush when the hinges are installed, as shown in photo 7-4. Unless you use an extra-large hinge, making the closer ¾ inch narrower than the mounting board will enable the inside edges to be flush, while allowing enough space to mount the hinges.

The mounting boards and the closers can be made from any thickness wood, although ¾ inch is best. A 1 × 6 will yield both a mounting board and a side closer from one length of wood.

Photo 7-4 — Mounting board and side closer should be flush on the inside edges.

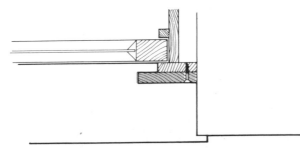

Illustration 7-5 — Mounting boards should be screwed to the window frame in an area where the screws won't interfere with the window.

With all four pieces cut, sand all sides and, using some type of shaping tool, round the edges a little. With the edges shaped smooth, the pieces are ready to be painted or stained. Do this before putting the pieces together.

When the pieces have been finished, fasten them together with spring-loaded cabinet hinges. The hinges should not be spaced more than 24 inches apart, and should be at least within 10 inches of each end of the side closers. Be sure the closer is about 1/16 inch up from the bottom of the mounting board to give it clearance to swing freely over the sill.

Before installing the side units on the wall, you should attach some type of knob or handle on each closer board. Don't position the knob or handle at the top or bottom of the closer; keep it somewhere near the middle.

The mounting boards must be fastened directly to the window frame, but in such a way that they do not interfere with the operation of the window. To do this, the mounting board should be screwed to the part of the window frame that you measured in chapter 3, "Sizing the Shade," to see if the shade could be mounted in this position. This area is shaded in illustration 7-5. Use 1½-inch # 8 flathead wood screws, so the screws will not penetrate the window channel or the sash weight channel if you make a mistake in screw placement. Put the outside edge of the mounting board directly against the window jamb, with the bottom of the unit resting on the windowsill. The difference in height between the mounting board and the closer should be at the top, where the closer must be shorter than the mounting board. The bottoms should be almost flush with the sill, leaving just enough room for the closer to swing freely.

With these pieces installed, you are ready to attach the fabric and complete the installation of your shade. Follow the instructions in the rest of this chapter.

ATTACHING THE FABRIC AND MYLAR

The remainder of this chapter pertains to both types of shades. Regardless of what type of roller box you have built, follow these instructions the rest of the way. All that remains is to hang the two pieces of fabric, attach the Mylar, make the bottom closer, run the pull cord through the shade, and test it out.

Before starting this part, look at blueprint sheet 5 to get an idea of what you will be doing. The fabric is fastened to the roller box through a series of snaps on the dowel rods and the fabric. There are a lot of snaps to put in place, and they have to be carefully positioned so the shade will be both efficient and versatile enough to change with the seasons.

As you can see from blueprint sheet 5, the snaps go on in two units. First, snaps are put on the dowel rods and the fabric. Second, snaps are put on the fabric and the Mylar. Each set of snaps must be positioned differently.

To attach snaps to the dowel rod, try to use a device that is a combination snap and screw. These are normally referred to as snap fastener studs. They are nothing more than fabric snaps with wood screws fastened in the middle, as shown in photo 7-5. This allows the screw to go into the dowel, firmly fastening the snap to the dowel, exactly where you want it. Snap fastener studs are often used for marine applications, so if your

Photo 7-5 — A snap fastener stud.

hardware store does not have them, try both marine supply and fabric stores. In a real pinch, you can make your own by putting a screw through the rivet hole in a snap and tightening it to the dowel.

Snaps are sold in a variety of sizes. Use large, heavy-duty snaps. These are normally referred to as # 5 snaps, heavy duty snaps, or type A snaps. The finished side of the snap should measure from ½ to ⅝ inch across.

Buy the snaps with screws in them first, and then get the other parts to match them.

Before putting the snaps on the dowel, you have to align the fabric and mark where both the dowel and fabric snaps will go. The first step is to position the piece of fabric that will face into the room. This fabric goes on the lower of the two dowels. To align these two properly, prop the roller box open and put a dowel in the roller box, on the bottom rod holder. Be sure you have the pulley on the side where you will want it on your finished shade.

Photo 7-6 — Mark the position for a snap by using the snap itself.

Working with the bottom roller, hold the end of the fabric that has seam tape on it against the dowel. Put one edge almost against the side of the pulley and thumbtack it in place. Position the other corner so the fabric hangs straight in the window. By lifting the free corner up or down, you will change the angle the shade hangs in the window. Position it so both sides of the fabric hang parallel to the side closers. Once in place, thumbtack it to the dowel. On wide windows, an additional tack in the middle of the dowel will help hold everything firm.

Carefully remove the dowel and the fabric from the roller box. On a firm bench, position the dowel with the pulley overhanging the end of the bench so the dowel lays flat. You now have to make a series of marks, on both the fabric and the dowel, for the positioning of the snaps to hold the two together. Be sure you don't disturb the thumbtacks holding the fabric in place. The

best way is to start at one end, measure in 1½ inches, and position a snap. Photo 7-6 shows the best way to mark the position for a snap. Put one edge of the snap against the seam tape, and use a pen to mark the fabric through the middle of the snap. Work your way across the fabric, spacing the snaps evenly, so the last snap is about 1½ inches in from the end of the fabric. Be sure you do not pull the fabric out of line while marking the snap placements. Snaps should be spaced every 12 to 16 inches.

With the marks made, DON'T unpin the fabric from the dowel. Go back over the fabric with a 6d finishing nail, hitting the nail through the fabric and lightly into the dowel at each snap placement mark. This will give you both a hole in the fabric and a starting pilot hole for the snap's screw. With this done, remove the fabric from the dowel.

Fasten the snaps to the dowel, using the pilot holes made by the nail. Blueprint sheet 5 shows which part of the snap to place on the dowel. Snap pieces are normally referred to as being male or female. You want the male part attached to the dowel.

With the snaps on the dowel, you now must attach the other half of the snap to the fabric. Be extremely careful you get the snap positioned facing the right way on the fabric. If you put it backward, the wrong side of the fabric will show. The mechanics of the roller mechanism of the shade are shown on blueprint sheet 6. Each piece of fabric and the Mylar must be rolled in a specific direction. Blueprint sheet 5 shows the direction the snaps must face on the front piece of fabric.

The snaps must go on the fabric so the finished side of the snap is on the inside of the fabric, and the operating part of the snap, the female side, is on the side of the fabric that you want facing the room. Use the holes that were punched through the material with the nail to position the snaps. Follow the directions on the snap package for installing snaps. Some snaps are fastened with a die and a punch, while others are attached using special pliers.

With the snaps on the front piece of fabric, and the front dowel rod, assemble the two and hang the dowel rod in the roller box. The shade should hang parallel to the side closers. If it all lines up, take the fabric and dowel rod down and repeat the positioning process with the top dowel rod and the back piece of fabric. Be sure you have the proper face of the fabric facing out the window when you align and pin the second piece of fabric to the dowel.

Mark the holes for the snaps, and attach the snaps to the dowel rod the same way.

Check the illustration on blueprint sheet 5 to see the difference in the direction the snaps must face. On the back piece of fabric, the snaps go on with the finished, smooth side of the snap on the side of the fabric that will face toward the middle of the shade, and the operating part of the snap, the female part, on the side that will face toward the outside of the window. Both pieces of installed fabric should have the operating side of the snap on the fabric face you want to see, and the smooth part of the snap on the fabric face that goes toward the middle.

If you look at blueprint sheet 6 you will get a better understanding of why this has to be. The mechanics of how the respective rollers have to roll are somewhat complicated, but they will only work one way. Follow the drawings exactly, and you'll have no problems. The Mylar must always be on the outside of the roll. When you change the position of the Mylar the direction the rollers turn must also change.

With both pieces of fabric attached to the dowels, put both dowels in the roller box and check their final alignment. If the two pieces of fabric line up, you're ready to proceed to the next step. If not, find out which piece of fabric does not align with the side closers, and what snaps have to be moved to make it align. You can drill out the rivet part of a snap to move it, but it is hard, and the chances of damaging the fabric are high. So take your time, and get the snaps in right during the initial alignment stage.

With both pieces aligned, you are now ready to attach the Mylar to the fabric.

First, work with the piece of fabric that faces out the window. Remove the fabric from the top dowel, and lay it flat on a work surface. Be sure the side that will face out the window faces down. Position the other piece of fabric on top of it, with the side that will face the room facing up just as though you had placed an entire shade flat on a table. The finished assembly should have the piece of fabric that faces out the window on the bottom and the piece of fabric that faces the room on top of it. The top edges of the two pieces should be staggered by the center-to-

Photo 7-7 — Lay out both pieces of fabric, and be sure they are staggered by the center-to-center distance.

center distance (on your data sheet), as shown in photo 7-7. This will enable you to mark one set of holes in both pieces of fabric at the same time to align them, exactly as they will be when hanging in the roller box. With these two pieces positioned at the top and along the sides, place a piece of Mylar over them.

The Mylar should be cut ½ inch narrower than the finished pieces of fabric. It should start 2 inches below the top of the front piece of fabric and be as long as the front piece of fabric at this time, although it will be trimmed to length later.

The Mylar is the only exotic type of material in the shade. It is also a very important part of the shade. You may be tempted to leave it out, as buying it is somewhat of a nuisance and it has to be ordered well in advance. But don't leave it out. We have used standard "space blankets" with good results, although they cost more than mail-ordered Mylar.

When the Mylar is aligned with the fabric, you have to mark where snaps will go on all three pieces. The snaps on all three pieces have to be in identical places, as the Mylar has to be interchangeable from one piece of fabric to the other. The best way to do that is to punch a hole through all three pieces at once. Be aware of potential damage to any underlying surface when you punch holes through the three layers.

Measure in from each edge of the Mylar 2 inches, and down from the top edge about ¾ inch, as shown on blueprint sheet 5. Make a punch with a nail or an awl through all three pieces of fabric. Working across the Mylar, continue to make punches on the same line, spacing your snaps 12 to 16 inches apart. A 30-inch-wide window should have at least four snaps.

With the holes for snaps marked through all three layers, first put snaps on the Mylar. Be very careful you get the snaps on the right sides of the fabric and the Mylar. If you get the Mylar turned around during the snap-fastening phase, it will end up out of sync with the two pieces of fabric and may not fit.

The best thing to do is to put the snaps on the Mylar while it is still in place on the other two pieces of fabric. The female part of the snap goes on the Mylar. This is the part of the snap with the finished, smooth backing piece. The smooth back of the snap should face the two other layers of fabric when installing the snap. Install the snaps on the Mylar before removing it from the pieces of fabric, and you won't get confused.

Next, put snaps on the piece of fabric that faces into the room. This piece gets the male part of the snap. The operating part of the snap should be on the back face of the fabric, and the flat fastening side on the side of the fabric that faces into the room. These snaps will not show when the shade is in operation; they will be hidden by the roller box. Refer to blueprint sheet 5 to be sure you have the snaps on the correct side of the fabric.

The snaps go on the back piece of the fabric with the flat mounting side facing toward the middle of the shade, and the operating side on the side of the fabric that faces out the window. This allows the Mylar to be hung on the outside of the fabric during the summer so that the Mylar faces out the window. In this mode it will reflect most solar heat directly out the window, keeping the room cool.

With all the snaps positioned on the fabric and Mylar, you are ready to trim everything to length.

Snap the two pieces of fabric to their dowel rods, and install the dowel rods in the roller box. Be sure you have the right fabric on the right dowel rod. Do not snap the Mylar to the fabric at this time.

Close and latch the front of the roller box. Pull the two pieces of fabric all the way down. The fabric will overhang the bottom of the window somewhat. Gently pull the fabric taut, and mark where the fabric meets the top of the windowsill. Make a light mark on both edges of both pieces of fabric at this point. The pencil marks should show where the fabric first touches the windowsill when pulled all the way down.

Take the fabric and dowel rods out of the roller box, and remove the fabric from the dowel rods. Work with one piece of fabric at a time from this point. Measure 3 inches from the mark on each side of the fabric toward the unfinished short edge of the fabric. This will give you the point where the fabric should be cut off. Cut the fabric so it is 3 inches longer than the window opening.

Following illustration 7-6 and photo 7-8, turn up a double 1-inch hem in the fabric and sew the hem. This extra thickness of fabric will be the bottom seal against the windowsill.

With a hem sewn all the way across the front piece of fabric, do the same to the back

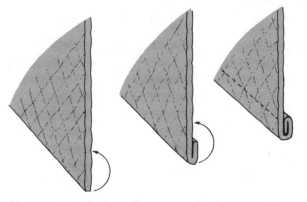

Illustration 7-6 — Turn up the hem and sew the bottom of the fabric like this.

Photo 7-8 — Turn up and carefully sew a hem all the way across the unfinished end of the fabric.

piece. Be sure to measure the fabric correctly. Don't cut the fabric short; the measurements will leave an extra inch of fabric for play.

With both pieces of fabric hemmed, you are ready to cut and attach the bottom molding. This is made of three pieces of wood. Illustration 7-7 shows the three wooden pieces and the fabric assembled. Note how the fabric overhangs the bottom of the combination by ¼ inch. This fabric serves as the seal when pressed against the windowsill.

The thickest of the three pieces of wood goes between the two pieces of fabric. This piece should be ¾ inch thick. The other two pieces are best made of ¼-inch lattice material. This is available at most lumberyards.

All three pieces should be cut between ¾ inch and 1 inch wide. The exact width does not matter. It is best to cut the center piece to whatever width of lattice you can find.

The length of the three pieces is found

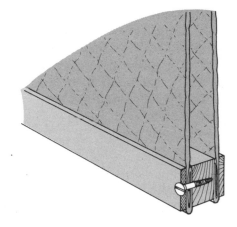

Illustration 7-7 — The bottom molding and fabric are screwed together.

by simply closing both side closers, and measuring the distance from the edge of one closer to the edge of the other closer. Measure to the inside edge of the closer, not the mounting board. From this figure, subtract 3 inches. The bottom molding pieces should be set back from the closers by 1½ inches on each side.

To attach the three bottom pieces, simply attach each piece of lattice to the center piece, using ¾-inch # 6 flathead wood screws, going right through the fabric.

The three bottom pieces should have about ¼ inch of the thick hem of each piece of fabric hanging out the bottom of the bottom molding. Pressed against the windowsill, this excess fabric will serve as a good barrier to air infiltration when the shade is pulled down.

The bottom pieces should be stained or painted before finally being attached to the fabric. You can also attach decorative knobs

to the bottom molding to aid you in pulling the shade into place.

With both pieces of fabric firmly attached to the bottom molding, cut the Mylar before installing the assembled parts in the roller box. Measure the distance on the front piece of fabric from the top of the bottom molding to the set of snaps used to hold the Mylar. Subtract 1 inch from that measurement, and that is the length of the Mylar when measured from the snaps on the Mylar. The overall length of the Mylar will be slightly longer. The end result should have the Mylar hanging between the two layers of fabric and ending about 1 inch above the top of the bottom molding.

Before you install the assembled fabric and dowel rods in the roller box, there is one more very important step. You must put rubber bands over the center of the pulleys, as shown in photo 7-9. Use ¼-inch-wide rubber bands that fit around the center of the pulley tightly. This will give a good pulling surface for the pull cord, to make sure the cord won't slip. We have made shades to fit entire sliding glass doors, and the pull cords did not slip off the rubber bands.

It is now time to install the two dowel rods, with the attached fabric and Mylar, in the roller box. With all the pieces in the roller box, all that remains is to attach the pull cord and the cord tensioner.

Look at blueprint sheet 6 to see how the cord must be routed in order for the shade to work. The cord path stays the same, regardless of whether the shade is set for winter or summer operation. The difference in how the

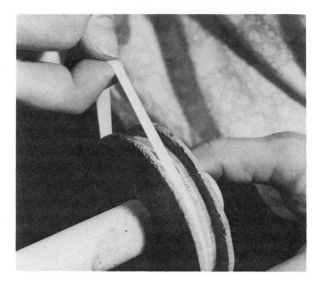

Photo 7-9 — Putting rubber bands over the center of a pulley gives the added traction the pull cord needs.

shade rolls is accounted for by the fact that you pull the front cord to roll up the shade when the Mylar is in the middle, and the rear cord when the Mylar is facing the outside. If you pull the wrong cord, the shade will slip and bunch up.

Route the cord as shown on blueprint sheet 6, and in illustration 7-8. Be sure you have it going the right way. You have to feed the cord through the cord hole in the bottom of the piece while the box is open.

We have found that standard drapery cord will serve your shade long and well. Some of the newer drapery cords are made of polyester-type cord and will slip more easily than the plain old cotton rope style.

The length of the cord is determined by where you want to attach your cord ten-

sioner. This is nothing more than a spring-loaded cord tensioner of the type used for draw drapes. There are two types made, one with a small cord passage hole at the bottom and another that will take larger cord. Get tensioners with as large an opening for the cord as you can find. The problem is that the electrical butt splice used to fasten the two ends of the cord together will not pass through the smaller openings of some cord tensioners. It will, however, pass through tensioners with large openings.

The draw cord must be cut to length and then spliced together to make an endless loop. The best way to fasten the cord together is to use a standard electrical butt splice. Working with standard drapery cord, a connector for # 10 – 12 wire will fit perfectly. Get uninsulated connectors from your hardware or electrical store. If your cord is slightly larger, the opening in the connector can be forced apart with an awl or other round device. Insert one end of the cord in one end of the connector and, making sure it is all the way in, crimp the connector tight on the cord. It is best to crimp the connector twice to get a good, tight fit. Repeat for the other end of the cord.

Don't install the tensioner and then try to cut the cord to fit that position. First, get a good idea of where you want the tensioner to go, then cut the cord to fit that position and use the butt splice to put the cord together. Then attach the tensioner. Tensioners are somewhat adjustable. Follow the directions with the tensioner to get the proper positioning for a tight pull cord.

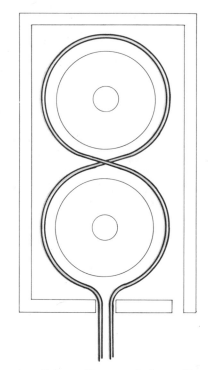

Illustration 7-8 — Route of the pull cord over the two pulleys.

OPERATING THE SHADE

With the cord fastened and the tensioner installed, you are ready to operate your shade. For winter operation, put the Mylar between the two layers of fabric, loosen the tensioner, and position the fabric so that it is fully extended, hanging over the entire window opening. Then move the cord so that the splice is on the back line of cord, just above the tensioner, as shown in photo 7-10. Reattach the tensioner.

With the cord splice located in this position, it will be a long time before you need to readjust the cord. Occasionally, the splice will work its way around the path of the cord until it gets in the way of the tensioner when you operate the shade. When this happens, you have to remove the tension from the cord and open up the roller box. Then maneuver the splice around so it is near the tensioner on the rear cord again with the shade pulled down. This will put it back as far as it can be from where it will get caught in the tensioner. There is always some slip and play in the cord, and as the shade is operated, the splice will again work its way around until it gets in the way.

When you switch the Mylar position for summer operation, you will have to move the splice so it is close to the tensioner, but this time it should be positioned on the front cord, as shown in photo 7-11.

With the splice properly positioned, close the roller box, latch it, and reconnect the tensioner. You are now ready to operate your insulating window shade. With the fabric pulled all the way down, and the pressure from the side closers removed by opening the side closers, pull the cord that is closest to you, and the shade will roll up. Here you should refer to blueprint sheet 6 to be sure you understand how to pull the cord for both winter and summer modes of operation. Roll it as high in the window as you want. When you stop, snap the side closers shut, and they will hold the shade in place. To close the shade, open the closers and, grabbing the bottom molding, pull the shade down until it

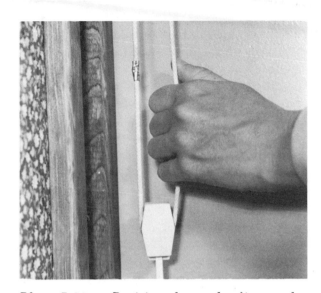

Photo 7-10 — Position the cord splice on the back cord, near the tensioner, for winter operation.

touches the windowsill and then shut the side closers.

To sum up the operation of the shade—in the winter the Mylar goes in between the two layers of fabric, and you pull the cord that is farthest from the wall. In the summer the Mylar goes on the outside layer of the shade, and you pull the cord that is closest to the wall.

That's it. You've built yourself an insulating window shade. From now on you'll be saving energy, and you'll be living in a more comfortable home. The best thing we can tell you to do at this point is to sit down and relax, and decide which window gets a shade next. Every window in your home should eventually have an insulating window shade.

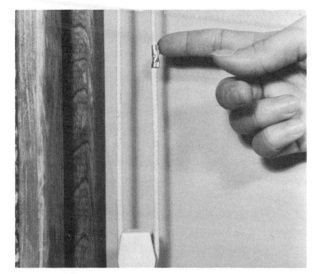

Photo 7-11 — Position the cord splice on the front cord, near the tensioner, for summer use.

Section III Blueprints

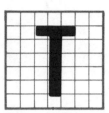his section may be the most unusual part of this book. The blueprints that are bound in after this page are designed for removal from the book. Each page is perforated. Tear the blueprint sheets out of the book one at a time, working carefully from one end to the other. Once all the pages are removed, refold them and store them in the envelope provided.

We feel that by keeping the blueprints with the book all the time, you'll be able to come back to the book years from now and reuse it. The envelope will protect the blueprints, yet allow you to take them to the shop to build your shade.

There are six blueprint sheets with Rodale's *Insulating Window Shade*. The package of blueprints should enable you to build either style of window shade.

Blueprint Sheet 1—Wall-Mounted Insulating Shade. This sheet details the pieces in a wall-mounted shade. Use the sheet to gain a complete understanding of the relationship between the various pieces in the shade.

Blueprint Sheet 2—Window-Mounted Insulating Shade. This sheet details the pieces in a window-mounted shade. Use this sheet only if you are building a window-mounted shade. Chapter 3 details the difference between a window- and a wall-mounted shade.

Blueprint Sheet 3—Wall-Mounted Shade Rod Holders. This sheet details every step in cutting the rod holders for a wall-mounted shade. The rod holders are the only pieces that might be considered tricky to cut, and this sheet gives complete instructions for measuring, marking, and cutting the rod holders.

Blueprint Sheet 4—Window-Mounted Shade Rod Holders. This sheet details every step in cutting the rod holders for a window-mounted shade. The rod holders are the only pieces that might be considered tricky to cut, and this sheet gives complete instructions for measuring, marking, and cutting the rod holders. This sheet should only be used for a window-mounted shade, as explained in chapter 3.

Blueprint Sheet 5—Fabric and Snap Placement. This sheet is used regardless of which type of shade you are building. It details exactly what snaps go where and how the different pieces of fabric relate to each other. This sheet should clear up any confusion concerning the assembly of these pieces. Pay close attention to it.

Blueprint Sheet 6—Winter/Summer Operation. This may be the single most important page in the book. It puts everything together to explain how a finished shade is to be used. You have to understand how the shade works before you install it. Read this sheet carefully, and be sure you know what goes on in an insulating window shade before you install and try to use your first shade. This sheet applies to both types of shades.

This completes your *Insulating Window Shade* book. We are confident that this book will fully enable you to build, install, and use insulating window shades. If you have any questions on the operation of the shade, or problems building a shade, please send all details to the Reader Service Department, Plans Books, Rodale Press, 33 East Minor Street, Emmaus, PA 18049. Please do not contact us on questions of material availability. We have printed all the names and addresses we have, in the materials section.